The Engineers

**A History of the Engineering Profession
in Britain, 1750-1914**

of related interest

The Professional Engineer in Society
Steve Collins, John Ghey and Graham Mills
ISBN 1 85302 501 1

The Engineers

A History of the Engineering Profession in Britain, 1750-1914

Jessica Kingsley Publishers
London

First published in 1989 by
Jessica Kingsley Publishers
13 Brunswick Centre
London WC1N 1AF

British Library Cataloguing In Publication Data
Buchanan, R. A. (Robert Angus), *1930-*
    The engineers : a history of the engineering
    profession in Britain, 1750-1914.
    1.. Great Britain. Engineering industries. Engineers,
    history
    I. Title
    331.7'62'000941

    ISBN 1-85302-036-2

Printed and bound in Great Britain by
Biddles Ltd, Guildford and King's Lynn

Contents

Preface

The engineering profession has made a large and distinguished contribution to British society over the past two centuries. It is a contribution, however, which has received little attention from historians apart from that devoted to the lives of a handful of the most notable engineers. This work is intended to remedy the deficiency by providing an overview of the origins, development, and ramifications of professional engineering in Britain during the period of intense industrialisation from 1750 to 1914. It is not primarily a biographical study of the engineers, although I have much to say about the leading engineers of whom it is possible to reconstruct an adequate personal profile. Nor is it an institutional history, even though I give a considerable amount of attention to the establishment and organisation of the engineering professional institutions in turn. What it does try to do is to pursue the notion of the engineers as a profession, treating individuals and institutions as they contribute to the elaboration of this concept, and trying to relate this to the wider context of eighteenth and nineteenth British society within which it occurred. There can be no doubt that it is a handicap to write about engineering without the technical expertise of an engineer, but it is the sort of handicap which an historian has frequently to face and seek to overcome. My studies, as an historian, of the history of engineering and technology, have brought me close to the achievements of the engineering profession over a long period of time, and I have consequently acquired a strong sympathy with engineers as people, responding to distinctive challenges and struggling to overcome complex technical problems. But lacking the technical expertise to understand the niceties of these challenges and problems, I have been unable to identify completely with the men I have studied. On the other hand, my experience in adjacent areas of historical study has enabled me to see the engineers in a broader context than they usually see themselves, and, I hope, to suggest relationships with other affairs of a social, political and economic character which have run parallel with their own preoccupations. There is a need for such an understanding of the engineering profession, and I have tried to fulfil it in the following pages.

This study has occupied much of my attention for a decade, and in that time I have incurred many debts, so that it is not possible for me to thank every individual who has helped me, much as I would like to do so. I am extremely grateful to all those engineers and institutional officers who have answered my queries, and particularly to:

Mr A. W. L. Naylor, Librarian to the Royal Aeronautical Society;
Mr T. J. Evans, Secretary to the Institution of Chemical Engineers;
Miss D. Bayley, Librarian to the Institution of Civil Engineers;

Mr J. Gurnsey, Librarian to the Institution of Electrical Engineers;
Mrs E. D. P. Symons, Archivist to the Institution of Electrical Engineers;
the Institution of Gas Engineers;
the Secretary to the Institution of Civil Engineers of Ireland;
the Iron and Steel Institute;
Mr H. Forsyth, Secretary to the Keighley Association of Engineers;
Mr J. A. King, Hon. Secretary to the Leeds Association of Engineers;
Mr I. J. Whittingham, Asst. Secretary to the Institute of Marine Engineers;
Mr S. G. Morrison, Librarian to the Institution of Mechanical Engineers;
Mr G. R. Strong, Secretary to the Federated Institution of Mining Engineers;
Miss Oblatt, of the Institution of Mining and Metallurgy;
Mr J. Wallis, Asst. Secretary to the Institution of Municipal Engineers;
the North-East Coast Institution of Engineers and Shipbuilders;
Mr R. Williams, Secretary to the North of England Institute of Mining and Mechanical Engineers;
Mr A. C. Wright, Deputy Secretary to the Royal Institution of Naval Architects;
Mr A. R. R. Goldsmith, Deputy Secretary to the Institution of Production Engineers;
Air Vice-Marshall S. M. Davidson, Secretary to the Institution of Electronic and Radio Engineers;
Mr J. B. Muirhead, Secretary to the Institution of Public Health Engineers;
the Institution of Engineers and Shipbuilders in Scotland;
Mr C. D. Morgan, Secretary to the Institution of Structural Engineers;
and the Institution of Water Engineers.

All these and others have responded helpfully to my requests for information. In fact, I only received one unfavourable response, and that is so unusual that it is worth mentioning: the Secretary to the Institution of Highway Engineers declared firmly that she thought my research was not worth while and refused to help. To all the others, however, I record my gratitude, even though I have not always been able to make use of the information with which they kindly supplied me.

I also give my thanks to Librarians, Archivists, and other keepers of records, amongst whom I particularly wish to mention:

the Keeper of Archives, Public Record Office, Kew;
the Director of the British Library;
the Director of Cambridge University Library;
Bodley's Librarian, Oxford University;
Mr L. Day, Librarian to the Science Museum, London;
the Director of the National Library of Scotland;
Mr M. Hewson, Director of the National Library of Ireland;
Mrs Susan M. Elkin, Asst. Librarian to the Science Library, the Queen's University of Belfast;
Mr R. F. Atkins, Director of Sheffield City Libraries;
the University Archivist, University of Glasgow;
Mr N. Higham, Librarian to the University of Bristol;
the Librarians at Newcastle City Library Local History Collection, Cardiff City Library, the Royal Cornish Institution Truro, Manchester City Library and Archives, Birmingham City Library and Archives, Liverpool City Archives and Local History Library, Leeds City Library, Bristol City Library, Bath City Library, Bolton Library, and Preston Library.

I am most especially grateful to Mr J. H. Lamble and his staff at the Library of the University of Bath for their support and patient assistance with a mass of inter-library loans. Amongst archivists and record offices, I am grateful to those at Tyne and Wear and Northumberland, both in Newcastle, and to Cardiff, Truro, Durham, Leeds and Bristol.

In its long process of gestation, several parts of this book have appeared in a preliminary shape in various books and articles, and I am grateful to the editors and publishers of these for permission to re-work this material in its complete form here. The most useful of these earlier accounts have been my articles:

'Steam and the Engineering Community in the Eighteenth Century' in *Trans. Newcomen Society* 50, 1978-9, 193-202;

'Gentlemen Engineers: the Making of a Profession' in *Victorian Studies* 26, 4, Summer 1983, 407-429;

'Institutional Proliferation in the British Engineering Profession, 1847-1914' in *Econ. Hist. Review* 2nd series, 38, 1, February 1985, 42-60;

'The Rise of Scientific Engineering in Britain' in *British Journal Hist. Sci.* 18,2,59, July 1985, 218-233;

and 'The Diaspora of British Engineering' in *Technology and Culture* 27, 3, July 1986, 501-524;

and my contributions to books:

'The British Canal Engineers: the men and their resources', in Per Sörbom (ed.), *Transport Technology and Social Change*, Tekniska Muscet Symposia, Stockholm, 1980, 67-68;

and 'Engineers and government in nineteenth-century Britain' in Roy MacLeod (ed.), *Government and Expertise* Cambridge University Press, 1988, 41-58.

Whilst I alone am responsible for the contents of this work, I was fortunate to have the help of a Research Assistant for three years, funded by the University of Bath, and I am grateful both to the University for this financial support, and to Dr J. Helen Bannatyne who assisted me from 1979 to 1982 with my basic research. I am also grateful to the Social Science Research Council, as it then was, for support with a travel grant to help Dr Bannatyne and myself get around institutions, archives, and libraries. I thank Mrs Nicola King and Mrs Aileen Mowry for exemplary typing services. And as always I am deeply grateful to my wife, Mrs Brenda J. Buchanan, for her scholarly support.

R. A. Buchanan
University of Bath
1st January 1989

Chapter One

Introduction: The Scope of Engineering History

Engineering, as an expression of the talent of *homo sapiens* for making artifacts, is very ancient. As a conception of professional status, however, conferring corporate consciousness on groups of its practitioners, it is relatively modern: the first fully professional engineers were products of the accelerating pace of industrialisation in the eighteenth century. About the middle of that century, a number of men engaged in practical works of construction and land drainage in Britain began to describe themselves as 'civil engineers'. The designation helped to establish a two-fold distinction. On the one hand, it differentiated those who adopted the title from the military engineers, still only partially organised in the Corps of Royal Engineers, but already performing vital service functions for the Army in relation to fortifications and the preparation of strategic roads and bridges. On the other hand, the style 'civil engineer' served to distinguish a man who had acquired some social standing from the large and incoherent body of men who regarded themselves as 'engineers' in the sense of being craftsmen skilled as millwrights, stonemasons, carpenters, blacksmiths and other assorted trades, who were employed in constructional tasks involving a measure of 'ingenuity' or enterprise. It is one of the tricks of the English language that such craftsmen and artisans have called themselves engineers, and continue to do so today in powerful trade unions which bear the title, while others have struggled to give the term a more precise definition by claiming for engineering the perquisites of being a 'profession'. This study is concerned only with the latter category of professional engineers, but it is important at the outset to recognise the ambivalence of the term. The historical fact is that the professional engineers in Britain have always coexisted with a much larger number of non-professional engineers, and that the latter have provided a substantial reservoir of manual skills and innovative talent from which the professional men have been readily recruited, so that the two groups have enjoyed a long but ambiguous relationship of common skills complicated by distinctions of social class.

Engineers as a whole have received very little attention from historians, and even professional engineers have been treated patchily, with little regard to their special skills and aspirations. Hitherto, the tendency has been to consider the development of engineering, in so far as it has been considered at all, as a series of biographies of the great engineers, with their subjects chosen almost exclusively from engineers who flourished between the mid-eighteenth and the mid-nineteenth centuries. Men such as John Smeaton and I. K. Brunel were certainly very significant both in advancing the frontiers of engineering competence and in providing models for the esteem of contemporaries and of posterity. But behind the great engineers there have been hundreds of little-known men who have contributed quantitatively much more to the profession than the handful of outstanding characters whose names are well known.

Moreover, the collaborative efforts of these engineers to express their professional aspirations through institutional organisations have been virtually ignored, even though studies of the organisation of other modern professions have demonstrated the historical value of such a perspective. There is a need, therefore, for a study which will do historical justice to the significance of professional institutions of engineers in Britain, and which will place these institutions and the engineers who composed them in the social context of the evolution of industrialisation since the middle of the eighteenth century.

The definition of engineering, for all the ambiguities which have been observed, is easy compared with that of professionalism. Traditionally, the professions were the skills of medicine, law and theology, which had been inculcated by the medieval universities, and subsequently by specialist institutions such as the Inns of Court in England. Each of these produced its own distinctive professional group, sometimes subdivided as in the division between the barristers and solicitors of the modern British legal system, with each maintaining its own entry requirements and code of conduct for its members. Any person wishing to practise as a medical doctor, a lawyer or a clergyman thus had to satisfy the appropriate group of his suitability and to subject himself to the discipline of the professional organisation. In addition to the specialisations within these traditional professions, the growth of industrialisation was accompanied by the emergence of new groups which aspired to achieve a similar professional status. These have included architects, surveyors, civil servants, teachers and engineers, and each group has endeavoured, with varying degrees of success, to achieve professional recognition for its members.

From the point of view of this study, professionalism is important because of the new significance which it achieved in the process of industrialisation as a social class giving shape and leadership to the processes of change. Marx recognised the existence of the professions but as they did not fit comfortably into his class system, differentiated according to sources of income, he did not develop any special role for them in his theory of social revolution.[1] Post-Marxian scholars were thus diverted from developing their understanding of the function of the professions until well into the twentieth century. However, Professor Harold Perkin has recently described them as 'the forgotten middle class', and has shown how the Industrial Revolution emancipated professional men from the dependence on wealthy patrons which characterised their earlier existence, and promoted them to become leading spokesmen for administrative reform and other aspects of social transformation in the industrialising process:

> 'What characterized the emancipated professional men as a class was their comparative aloofness from the struggle for income... Their ideal society was a functional one, based on expertise and selection by merit...'[2]

Perkin has defined the professional classes in terms of a distinctive social role rather than of access to wealth or income. That social role is one of effective - although not necessarily nominal - leadership, and it is derived from possession of a special education or expertise which, because it is valued by society, is rewarded by payments commensurate with the service of leadership. This functional definition resembles that of Samuel

Coleridge's 'clerisy', who 'comprehended the learned of all denominations'[3] or, much later, that of James Burnham's managers in his seminal study of the 1940s, *The Managerial Revolution*, in which a thoughtful Marxist tried realistically to account for the undoubted fact that the social revolution prophesied by Marx for the advanced industrial countries had not occurred, and no longer seemed likely to occur.[4] From all points on the political compass, in short, professionalism has come to be regarded as one of the significant formative developments of industrialisation.

Neither this recognition nor the increasing literature on the subject makes it any easier to provide a simple definition of professionalism. Carr-Saunders and Wilson, in their study of 1933, avoided giving any such definition, although they fell back when they needed to do so on the OED definition of 'profession' as: 'a vocation to which a professed knowledge of some department of learning or science is used in its application to the affairs of others or in the practice of an art founded upon it.'[5] Lewis and Maude, writing on *Professional People* in 1952 agreed that: 'This definition will certainly cover many, if not all, cases; but it may cover too much'.[6] They pointed out that, as it stood, it would include 'craftsmen of the higher sort' but not administrators or managers, and they preferred the definition offered by Carr-Saunders in 1949, in a Report on Education for Commerce, which described a profession as: 'any body of persons using a common technique who form an association, the purpose of which is to test the competence in the technique by means of examination'.[7] But they went on to note the distorting effect of placing all the emphasis on organisation: 'the history of professions is certainly more than a history of guilds' - and they referred in particular to the objection to this definition made by the senior engineering institutions, concerned with confining the pretensions of the new radio and production engineers, and stressing that a professional association must give training in a fundamental department of science, rather than a secondary or peripheral field.

Lewis and Maude also summarised with approval the more comprehensive definition of a professional group provided by an eminent engineer, Mr. W. E. Wickenden, speaking as president of the Institution of Electrical Engineers in 1950. Wickenden listed six dominant attributes conferring professional character on a group: possession of a body of knowledge or art; an educational process; a standard of professional qualifications; a standard of conduct; formal recognition of status by colleagues or the state; and an organisation devoted to common advancement and social duty rather than the maintenance of an economic monopoly. He concluded: 'Professional status is therefore an implied contract: to serve society over and beyond all specific duty to client or employer in consideration of the privileges and protection society extends to the profession'.[8] While all these attributes provide useful clues to understanding the nature of professionalism, they all beg important questions in terms of precision of definition: Wickenden's first attribute of 'a body of knowledge or art', for instance, fails to help in making distinctions between a 'professional engineer' and an 'engineering technician' of the type which has been widely accepted in countries other than Britain.[9]

Impressive imprecision thus surrounds the concept of professionalism, and this has not been dispersed by other recent studies, even when, like W. J. Reader's monograph of 1966, they recognise the importance of the subject. Like Perkin, Reader emphasised the functional role of professionalism in industrialisation: 'Professional associations multiplied to meet the growing complexity of Victorian industrial society and the advance of technology and science'.[10] He quoted with approval H. Byerley Thomson, writing in 1857 on *The Choice of a Profession*:

'The importance of the professions and the professional classes can hardly be overrated, they form the head of the great English middle class, maintain its tone of independence, keep up to the mark its standard of morality, and direct its intelligence.'[11]

But Reader failed to improve on earlier definitions on the subject.

Many writers on professionalism have paid tribute to the words of Francis Bacon:

'I hold every man a debtor to his profession, from the which as men of course do seek to receive countenance and profit, so ought they of duty to endeavour themselves by way of amends, to help and ornament thereunto.'[12]

The strong implication in these words of the moral responsibility of professional people to observe a code of behaviour and thereby to 'help and ornament' the status of their group is clear in most attempts to define the quality of a profession. Professional conduct, rigorously maintained by an organisation comprising the acknowledged leaders of the group, became a common feature of all the new professional associations as well as the old ones. A continuing procession of institutional foundations demonstrated this development in the nineteenth century. For engineers, the key date was 1818, when the Institution of Civil Engineers was created, and we will have much more to say about that event and the subsequent institutional proliferation amongst engineers. In the same period, the Institute of Architects was founded in 1837,[13] the Institution of Surveyors in 1868,[14] the Institute of Bankers in 1879,[15] and many other groups conformed to this pattern. Such bodies varied in their emphasis in placing more or less stress on instruction, or examination, or practical experience than others. But they all set out to maintain professional standards, adopting an ethical concept of the relationship between the professional man and his client. They were definitely not wage or salary negotiating bodies, although these thrived also in the same period, and tend to complicate the picture: the British Medical Association, formed in 1856, has always taken a lively interest in the pay of doctors and has thus failed to qualify as a fully fledged professional body. For most professional purposes, appropriate remuneration arose from a recognition of high quality service, reinforced by an important degree of monopoly control in dispensing this service, but the actual terms of payment have been subject to determination between the professional man and his client. Here, as elsewhere, in the tricky task of assessing grades of social distinction in Britain, the elusive characteristics of 'gentility' and 'gentlemanly conduct' were more significant than any specific system of bargaining.[16]

None of the general studies of professionalism mentioned so far has dealt particularly with engineering, although Carr-Saunders and Wilson gave an excellent summary of the development of engineering institutions[17] and the others commented on the importance of professionalism in engineering. Reader, for example, observed perceptively:

'Without the engineer, industrial England could not be. How, then, did he stand in English professional society?... The answer is: not very high.'[18]

He saw the social status of the engineer being retarded by what were frequently the obscure social origins of the engineers, and being frustrated by the 'heavy conservative bias of English society' in the mid-nineteenth century. This observation anticipated the thesis recently expounded by Martin Wiener on the decline of the industrial spirit in Britain after the middle of the nineteenth century, and it is a theme to which we shall return.[19] At this point, however, it is sufficient to note that it does not add to or detract from the emerging professional consciousness of engineers, who participated fully in the increasingly significant function of the middling classes in nineteenth century Britain in supplying leadership to society in the shape of vital expertise.

A modern sociological study which has concentrated on the professionalisation of one group of engineers is that by Gerstl and Hutton on the British mechanical engineers published in 1966.[20] Accepting the imprecise meaning of 'profession', they suggested that it contained three elements: specialised qualifications, independence in the establishment of its own code of conduct, and a strong commitment or attachment to intrinsic occupational goals. While these categories beg the same questions as those considered by Lewis and Maude, they are sufficiently clear to provide useful guidelines, while preserving a tolerant flexibility in our understanding of the concept.

In summary, therefore, professionalism may be regarded essentially as the institutionalisation of a social distinction - a distinction between the leaders and the led in which a particular expertise, acquired through approved methods of training and practised under the discipline of an accepted code of conduct, is put at the disposal of society and rewarded in terms of status and with remuneration which befits that status.[21] The privileges accruing to status vary over time and from one professional group to another, but their important feature is that they are acceptable to the leaders of each professional group, so that their acquisition through a professional career imparts fulfilment and even honour. The professional engineers were the leaders of a substantial but amorphous army of people who made a livelihood by exploiting talents of design and craftsmanship in the British Industrial Revolution. The fact that they rose to the top of this social pyramid demonstrates their mastery of the expertise involved, and makes them an exceptionally interesting group to study in a period when their skills were in great demand. But the very open-endedness of the profession towards the mass of skilled and semi-skilled technicians below them created tensions within the profession and threatened the security of the leaders. Professionalism thus came to possess an important dimension of social security, by which the leaders sought to buttress their position against erosion from below. This was made clear in the engineering lock-out of 1851, when the vitriolic letters attacking the trade unions in *The Times* were written by the son of Sir William Fairbairn under the pseudonym 'Amicus'.[22]

To the early British civil engineers, the goal of professional recognition must have seemed remote and daunting. They were, after all, a motley crew of men from a wide variety of practical backgrounds, but with little social standing in an hierarchical society which attached great importance to distinctions based on possessions, occupations, and family connections. They had no experience of a common education, or even of any formal school or university training. The fact that they began to think of themselves as professional men is attributable largely to the vision of John Smeaton, who was unusual amongst the early engineers in that he was the son of a Leeds attorney and had already established his reputation as a scientific investigator of some distinction before becoming a civil engineer. Once set on the road to professionalisation, however, the British engineers achieved it remarkably quickly, and an outstanding reason for this success was the exceptional quality of the men who emerged as leaders of the new profession.

Men like Smeaton, Watt, Rennie, Telford, the Stephensons, and the Brunels, became folk-heroes of the early Victorian Age, and their reputations did much to ensure the prestige of their profession. The enormous popularity of this generation of engineers created a demand for biographical studies, and foremost amongst these was Samuel Smiles' *Lives of the Engineers*, first published in 1862. Smiles himself was not an engineer; nor, for that matter, was he a trained historian. He was a medical man turned journalist and self-appointed prophet of the gospel of 'self-help' - the view that it was possible and desirable to improve one's position in life by exercising the sterling qualities of independence, hard work, and integrity. In this objective, Smiles mirrored the dominant middle class 'laissez-faire' attitudes of the society in which he lived, and it was his ability to give firm shape and historical demonstration to this conventional wisdom of mid-nineteenth century Britain that made his work so popular.[23]

The formula devised by Smiles was that of argument by cumulative biographical anecdote. By piling up one potted biographical success story of self-help upon another he produced an impressive parade of public figures who had made good for themselves in the world. His first big success came in 1859, with the publication of *Self Help*, but he had already come to recognise that the engineers of the eighteenth and nineteenth centuries in Britain provided him with excellent illustrative material for his thesis, and he had produced a brief biography of George Stephenson as an example of exceptional ability manifesting itself in a man of humble origins, who overcame all adversity by means of his genius and determination to prove to the world that the future lay with locomotive-powered railways. This essay was then incorporated into a three-volume survey of British engineers, in which the first volume dealt with the pioneers and with the career of James Brindley; the second covered the lives of John Smeaton, John Rennie, and Thomas Telford; and the third dealt with George and Robert Stephenson.

For all their didactic intention, these volumes, published as *Lives of the Engineers*, were and remain excellent biographical studies. Smiles was thorough in examining the documentary sources available to him, and he was not slow to take advantage of physical evidence in the shape of engineering monuments or of oral evidence in the form of reminiscences about the subjects of his research. There seems little doubt that he rescued material which would have been lost without his display of interest. He approached his individual subjects with sympathy and sensitivity, and he succeeded in presenting the profession of engineering to a wider public than it had enjoyed before. So successful was Smiles, indeed, that contemporary engineers hoped to have the graceful charm of his pen applied to their own careers. In the case of James Nasmyth, for instance, Smiles was persuaded by the inventor of the steam hammer to re-write his autobiographical notes and to present them to the public for him. Usually, however, Smiles preferred to deal with subjects who were already dead, and for his more substantial engineering biographies he chose to work on James Watt and Matthew Boulton.[24]

Samuel Smiles did much to popularise the engineers about whom he wrote, but he did not create a fashion which others were quick to follow. In fact, there is a dearth of good British engineering biography between the 1860s and the 1950s which, from the point of view of providing an account of the development of the engineering profession, causes a remarkable irregularity in the quantity and quality of the biographical material available. True, there was some filling in of the gaps in the pre-1860 generation with biographies of Sir William Fairbairn[25] and Charles Blacker Vignoles[26], and Thomas Mackay wrote a *Life* of Sir John Fowler when that engineer died in 1898, but this is a dull and disappointing book, except where it is enlivened by direct quotations from Fowler's correspondence.[27] Other prominent late-nineteenth century engineers such as Sir Joseph Bazalgette, Sir John Coode, James Mansergh, and Sir Benjamin Baker, have received no worthy biographical treatment. It would seem that what is lacking in this post-Smiles period is not so much material for biography as biographers to take up the challenge of writing on the subject with the same verve as Smiles.

Another curious fact about the post-Smiles period (that is, the period after the publication of *Lives of the Engineers* - Smiles himself lived on to 1905) is that as far as any studies were made in engineering biography, they tended to concentrate on figures in the earlier generation already explored by Smiles rather than on engineers who flourished after 1860. Thus, the series of sound biographies by those pioneers of the Newcomen Society for the History of Engineering and Technology, H.W. Dickinson, Rhys Jenkins, and Arthur Titley, chose as their subjects Watt, Boulton, and Richard Trevithick.[28] Likewise, Sir Alexander Gibb wrote a vigorous biography, *The Story of Telford*, in 1935, and Lady Celia Noble wrote her charming study, *The Brunels - Father and Son*, in 1938. The Brunels had, of course, been the subjects of earlier studies: Sir Marc Brunel in the *Memoir* by his Assistant Engineer Richard Beamish in 1862, and I. K. Brunel by his son in the biography published in 1870.[29] These are painstaking but somewhat undistinguished works, failing to do justice to their colourful subjects. So colourful, indeed, were the careers of the Brunels, that it is a mystery how Samuel Smiles failed to write about either of them. It has been suggested that Smiles did not see the younger Brunel as a model self-made man, but in this respect Brunel had exactly the same advantage of being the son of an eminent engineer as that enjoyed by his great contemporary Robert Stephenson (they both died in 1859), who was certainly not excluded from the Smiles pantheon on that account.

The relative dearth of engineering biography in the post-1860 period was qualified by a few autobiographical studies, although some of these were only published many years after the death of their subjects. The works of Joseph Mitchell, John Brunton, and Sir Daniel Gooch, are all of interest in this respect. Mitchell had been the son of Telford's supervisor of road construction in the Highlands of Scotland, and when his father died suddenly in 1824 Telford gave the responsibility to the young Mitchell who was thus launched onto a long engineering career building roads and railways in Scotland. His *Remi-*

niscences first appeared in 1883 in two volumes, but was originally published privately, apparently because of the caustic comments which he made about surviving colleagues, and the book has never been widely recognised for the valuable insights which it gives into the professional life of a nineteenth century engineer. For example, Mitchell had attended early meetings of the Institution of Civil Engineers while living in Telford's house in the early 1820s, and was able to give an account of the way in which they were conducted.[30]

John Brunton's Book was originally an autobiographical manuscript prepared by its author, the son of the pioneer locomotive engineer William Brunton, for the entertainment of his grandchildren. It was discovered and rescued from obscurity by the economic historian Sir John Clapham, for whom it was published by Cambridge University Press in 1939. It is a discursive autobiographical fragment, describing Brunton's early life on railways in Britain and his subsequent career in Turkey, where he acted as Assistant Engineer to I.K. Brunel on the Renkioi Hospital during the Crimean War, and in India, where he undertook several railway projects as an agent of the British Raj.[31]

The most substantial of the autobiographical works by late-nineteenth century engineers was that written by Sir Daniel Gooch - *Memoirs and Diary* - which gives an illuminating account of his work on the Great Western Railway and the Transatlantic Telegraph.[32] Although a version of these papers was published in 1892, they did not become generally available until 1972, after appearing in a family sale at Sothebys in 1969. It is indicative of the loss of public interest in the lives of engineers after 1860 that the work of even such a distinguished and honoured engineer as Gooch (he was made a Baronet in 1867) should have failed to attract the attention of a competent biographer.

Engineering biography was revitalised in the 1950s largely as a result of the work of L. T. C. Rolt, who wrote a succession of vivid studies which revived the Smiles technique, but without the didactic and moralistic quality which, in retrospect, mars so much of Smiles' work. It is significant, however, that Rolt returned for his major subjects to the pre-1860 period which Smiles had so brilliantly exploited: *Isambard Kingdom Brunel* (1957), *Thomas Telford* (1958), and *George and Robert Stephenson* (1960). He also wrote on James Watt and Thomas Newcomen and did some shorter biographical studies, but they were virtually without exception devoted to what has seemed to become increasingly an Heroic Age before 1860.[33] Rolt stimulated a considerable burst of activity in engineering biography by other authors, although few of these newer works have achieved the quality of animated readability which is such an outstanding characteristic of Rolt's books. Three biographical symposia have been particularly notable in generating close study of the works of the great engineers: *The Works of Isambard Kingdom Brunel* (1976), edited by Sir Alfred Pugsley; *Thomas Telford: Engineer* (1980), edited by Alistair Penfold; and *John Smeaton FRS* (1981), edited by Professor A. W. Skempton. Again, it will be observed that all these post-1950 studies have gone back to the Heroic Age for their subjects, leaving in an historical limbo the lives of the engineers of succeeding periods.

It is not easy to account for this concentration on pre-1860 engineering biography. It could be argued that none of the post-1860 engineers were from the same heroic mould as their great predecessors, although this would be difficult to maintain in view of the dramatic achievements of men such as Fowler, Gooch, Armstrong and Whitworth. Alternatively, it could be argued that these men, however great their talents, had less opportunity for innovation than their predecessors so that their works were consequently less interesting, and their achievements primarily a matter of managerial organisation and entrepreneurship. But such an argument neglects the excitement of novelty in electrical, automobile, and aeronautical engineering, and it also overlooks the organisational aspects of canal and railway building, so that the implied contrast is by no means as sharp as may be supposed, even though the trend towards increasing specialisation usually confined engineering practitioners to a narrower range of activities than those of polymaths like I. K. Brunel or Robert Stephenson. Perhaps the most plausible explanation is that the development of professional institutions amongst engineers in the second half of the nineteenth century made the contributions of individual engineers appear less important than in the earlier period, when there had been fewer engineers and a narrower range of precedents available to provide a basis for engineering judgments. It was this openness of professional opportunities before 1860 which called for the exercise of talents on an heroic scale, and which made those who were able to display these talents appear as giants to posterity. But one unfortunate result of concentrating biographical attention on the pre-1860 period has been to leave a serious gap in our knowledge of individual engineers since that time and thus to distort the way in which the profession has been regarded.

In its most extreme manifestations, the distortion of engineering biographical history has resulted in a form of hero-worship. Although the treatment of the lives of the engineers in the period before 1860 as most elegantly expressed in the works of Smiles and Rolt was always robust, so that they did not fail to recognise small warts in their portraiture, it does not completely escape from such hero-worship. Thus, although Smiles did not really warm to the aloof character of John Rennie, he nevertheless gave him full credit for the Bell Rock Lighthouse, which has subsequently been claimed (rightly) for Robert Stevenson, founder of the Scottish dynasty of lighthouse engineers.[34] And although Rolt clearly found some aspects of Telford's character opaque and puzzling, he still gave him all the credit for canal schemes which should properly have been shared with William Jessop. But then Jessop was one of those unfortunate engineers of the Heroic Age who only rose to a footnote in Smiles, so that he has only recently received his due in the biography by Hadfield and Skempton.[35] In other cases - the Stephensons, for instance, with Smiles, and I. K. Brunel with Rolt - the admiration of the biographer for his subject is virtually unqualified and has probably been responsible for a degree of over-adulation by subsequent historians of engineering. To be more specific, Rolt's need to account for the break-down of Brunel's last epic project - the construction of the steam ship *Great Eastern* - led him to elaborate an ingenious thesis which placed most of the blame on the disloyalty of Brunel's colleague in this venture, John Scott Russell. Recently, George S. Emmerson has demonstrated that this argument can no longer be satisfactorily sustained.[36] Even though the break-down of the *Great Eastern* enterprise remains an intriguing problem in engineering history, the available evidence does not justify making Russell alone the scape-goat.

This tendency to hero-worship has thus been responsible for some distortions in our understanding of the careers of the leading engineers from the Heroic Age before 1860.

More serious, however, as a reflection on the historiography of British engineering, has been the remarkable lack of interest in any engineers except those in the very front rank. For the pre-1860 period, the lives of some of the more obscure engineers are momentarily illuminated as they pass through the spotlight focussed on the Great Men: Watt's engine builders, Telford's Resident Engineers on the Caledonian Canal, George Stephenson's rivals amongst early locomotive men, and I. K. Brunel's Assistant Engineers on the Great Western Railway, all appear briefly on the biographical stage in strictly subordinate roles, and it is almost impossible to find out any details about the careers of these lesser men once they have passed out of their association with the heroic figures. And for the period after 1860, even the more prominent engineers fall into this penumbra area of relative obscurity. This raises an important question for our study of the engineering profession: what was the quality of this anonymous army of engineers?

In attempting to suggest a way in which this question may be answered, it is important first to recognise the size of the army of engineers. In 1850, the combined membership of the Institution of Civil Engineers and the newly formed Institution of Mechanical Engineers was about 900. Forty years later, in 1890, there were eight more national engineering institutions, and the total membership was about 15,000. In 1910, with seventeen national institutions, the total number of engineers enrolled had risen to about 36,300.[37] These figures demonstrate a substantial increase in the number of engineers over the period of this study. In total, they can hardly have numbered less than 50,000, even allowing for a large number of long-term enrolments amongst institution members. This is a substantial constituency, the analysis of which must inevitably be perfunctory in many respects, even though it should be possible to make some useful generalisations about the composition of the body concerned.

One way of piercing the obscurity concerning the personalities of these 50,000 British engineers in the nineteenth century is through the obituary notices which appeared in the journals of the national engineering institutions and other organs of the engineering press. Fortunately, this task has been assisted by the publication in 1975 of a *Biographical Index* compiled by S. P. Bell on the basis of a survey of forty contemporary engineering journals (only twenty of which carried regular obituary notices) for British engineers who died in or before the year 1900.[38] The first thing that must be observed about this compilation is that, with references to some 3,500 engineers, the list represents only a fraction of the total, covering less than one tenth, so that it leaves the great majority of the professional engineers of the period shrouded in anonymity. Secondly, Bell's compilation is essentially nothing more than a list of references to obituary notices, so that, for all its utility as an index, it does not itself provide much basis for biographical assessments. But it does give names, titles, and dates for the engineers listed, so that it is possible to make a few simple generalisations. It can be calculated, for instance, that ninety-seven of the 3,500 listed engineers possess honorific titles. This proportion of one in thirty-six does not seem excessively generous. Nevertheless, for a new profession, winning public esteem for itself, it is not insignificant, particularly when it is remembered that some of the pre-1860 generation declined British honours, including Telford (although he did accept a Swedish knighthood), Rennie (although his son accepted a knighthood on the completion of London Bridge) and George Stephenson (although he did receive a Belgian honour).

With this reservation, it would seem reasonable to regard the ninety-seven titled men as representing the *crème de la crème* of the British engineering profession in the nineteenth century. Who were they? Twelve were peers, and it seems likely that most of these, like the Dukes of Devonshire and Sutherland, were only 'honorary' engineers, although sufficiently sympathetic to the profession to receive obituary notices on their deaths. The dozen include at least one, Lord Armstrong, who rose through the ranks (although he began his career in Newcastle as a solicitor) of the profession as an entrepreneurial engineer. The Army accounted for fifteen of the ninety-seven titled engineers, and it seems probable that these were mainly, if not entirely, members of the Corps of Royal Engineers like Lt. General Sir Charles Pasley and Lt. General Sir William Denison, although only three are specifically credited with membership. The Navy was responsible for four: Admiral Sir Henry Denham, Admiral Sir Robert Robinson, Admiral Sir George Richards, and Rear Admiral Sir Charles Parry. It is not evident from the summary list whether or not they had worked their way through the Senior Service as Engineer Officers, but given the way in which engineering was regarded in the Navy before 1914 this seems extremely unlikely. Of the total list, only seven appear to be credited with baronetcies, but almost certainly some to whom knighthoods are attributed should come into this category. The general pattern, as far as the civilian engineers were concerned, seems to be one in which knighthoods were granted to practising engineers for some specific distinction, such as that awarded to Sir Marc Isambard Brunel when the Thames Tunnel approached completion in 1841, and to Sir Thomas Bouch on the opening of the ill-fated Tay Railway Bridge in 1877. A few then acquired the hereditary title of a baronetcy as a further reward, as when Sir John Fowler received his baronetcy in 1890 on completion of the Forth Railway Bridge, having already had the accolade bestowed on him in 1885 as a result of his services to the government in Egyptian affairs.[39]

These observations demonstrate that the engineering profession had achieved a considerable degree of public acceptance and praise by the end of the nineteenth century. Public honours are not themselves matters of engineering significance, and as we have noticed some of the most distinguished engineers of the century deliberately refused them. But they do give a guide to professional credibility, and the fact that a small but impressive proportion of the total number of engineers received recognition in this way says much about the standing of the profession as well as about the quality of the individuals concerned.

In making a preliminary assessment of the role of professional engineers in British society, it is pertinent to refer to that monument of prosopographical research, the *Dictionary of National Biography*. This massive enterprise was founded in 1882 and was completed in 1903 when the first part of the *Concise DNB* appeared.[40] The total number of biographies recorded is 30,378, of which some 489 can be considered as dealing with engineers, so that at first glance the proportion of one in sixty-two allocated to engineers does not seem over-generous. But the proprietors and editors aimed at catholicity, so that the great political and aristocratic names of British history jostle in strict alphabetical order with a remarkable range of Scottish kings, Welsh saints, and Irish poets; with highwaymen, actresses, and murderers; with nonconformist divines, Jesuits, and regicides; with engravers, publishers, and antiquarians. Hopkin Hopkin receives an entry as a famous dwarf; Ruth Osborne is memorialised as the last woman to be executed as a witch by ducking; and John Pace is recorded as a professional fool. The rich tapestry of British history is amply represented in these pages, and if the selection of entries reflects the middle class intellectual bias of the compilers and the conventional wisdom of late Victorian Britain, that it only to be expected. The surprising thing, indeed, is not that so few engineers are mentioned, but rather that they receive such substantial representation in a selection which is so widely drawn.

The identification of 489 entries as engineers has been made by an analysis of the *Concise DNB*. The possibility of some marginal error in this figure must be recognised, because the careers of some engineers may have been obscured by other aspects of their lives which minimise or even eliminate reference to engineering in the summary entry on them. Thus, for example, James Douglas (1753-1819) is described as 'divine, antiquary, and artist', although he also obtained a commission in the Corps of Engineers, in which capacity he worked on the Chatham Lines and developed the interest in archaeological excavation for which he is mainly remembered today.[41] It has not been possible to determine how many examples of such occlusion occur in the *Concise DNB*, but it is probably not large.

The total of 489 entries can be conveniently broken down into four categories. The smallest of these is a group of twenty-nine classified as belonging to the period before that with which this study is primarily concerned. The earliest such entry is that for Peter de Colechurch (d.1205), to whom is attributed the construction of the first stone bridge over the Thames in London in 1176 (he is also described as an architect rather than an engineer, and as chaplain to St. Mary Colechurch). There are several Tudor and Stuart fortification engineers listed, and also such proto-engineers as Sir Hugh Myddelton (1560?-1631), projector of the New River Company; Sir Cornelius Vermuyden (1595?-1683?), land drainage engineer; Thomas Savery (1650?-1715) and Thomas Newcomen (1663-1729), pioneers of steam technology; and Field Marshal George Wade (1671-1748), builder of roads in the Highlands of Scotland. All persons who died before 1750 have been grouped in this category.

The rest of the entries have been divided into one general and two special categories. The special categories are military engineers and professional engineers. There are 137 entries identifiable as military engineers. The great majority of these are members of the Corps of Royal Engineers, and most of them are men of distinguished rank and service. They include engineers like General Sir Arthur Cotton (1803-1899), expert in Indian irrigation; Lt. General Sir William Denison (1804-1871), who performed outstanding public services in Australia and India; and General Sir Charles Pasley (1780-1861), the first director of the military engineering college at Chatham and one of the first government inspectors of railways. Members of the Royal Artillery have generally not been included in this category, even though many military engineers trained at the Royal Military College in Woolwich.

The professional engineers constitute a category numbering 129. These are the men from whose entries it can be deduced that they were members of engineering professional institutions or very closely associated with such institutions. Thus James Brindley (1716-1772) has not been included on the grounds that his career preceded the formation of professional organisations, whereas George Stephenson (1781-1848) has been included because despite his robust criticism of professional bodies he did eventually accept the presidency of the Institution of Mechanical Engineers. Entries in the *Concise DNB* usually mention membership of the Institution of Civil Engineers where appropriate, but other institutions are only rarely mentioned so that it is possible that some engineers who have been relegated to the 'general' category deserve to be considered as 'professionals'. The selection contains the main founding fathers of the profession such as John Smeaton (1724-1792), Robert Mylne (1734- 1811), Thomas Telford (1757-1834), and John Rennie (1761-1821). It includes dynasties of engineers such as the Brunels, the Stephensons, and the Stevensons. Later nineteenth century engineers are less prominent, but they include Sir John Fowler (1817-1898), William Froude (1810-1879), and Thomas Hawksley (1807-1893). One difficulty here is that several important engineers of the period such as Sir Charles Parsons and Sir Benjamin Baker survived into the twentieth century so that their entries must be sought in subsequent volumes of the *DNB*. There are some notable omissions from the compilation: William Jessop (1745-1814), an early Smeatonian; H. R. Palmer (1795-1844), founder-member of the Civils; and Joseph Mitchell (1803-1883), Scottish road and railway builder, are amongst those whose absence seems unjustified. But for the most part, a good selection of leading engineers, mainly associated with the Civils, receives recognition.

The final category consists of 194 engineers in a fairly general sense, including inventors, 'mechanicians', instrument makers, 'electricians', mining engineers, and possessors of other partially specified skills of an engineering nature. It is likely that a proportion of these deserve promotion to the 'professional' category, but it is useful to establish that a substantial number of the men who were seen as practising engineers in the nineteenth century were far from being professional men in the sense that Smeaton, Telford, and I. K. Brunel understood the role of the engineer. As far as possible, architects and surveyors have been excluded from this category, as have men who were primarily entrepreneurs or manufacturers of engineering products, or contractors for engineering works. There remains, however, an impressive range of non-professional engineers who received *DNB* notice, including Joseph Bramah (1748-1814), inventor and locksmith; Sir George Grove (1820-1900), remembered as a writer on music even though he was articled as a civil engineer and became a member of the Civils in 1839; and Lionel Lukin (1742-1834), to whom is attributed the invention of the 'unsubmergible' lifeboat.

It is interesting to observe that, excluding those who flourished before the middle of the eighteenth century, the men selected for recognition as engineers by the *DNB* fall into three roughly equal categories of military, professional, and general engineers, although it would be unwise to attach any great significance to this circumstance. Of the military engineers, it is probably true to say that the majority of those mentioned were recognised primarily for their public service as colonial governors, administrators, and inspectors, although many of them were very able engineers and undertook major works of irrigation, water control, and construction. They were, after all, the only engineers to receive any sort of formal instruction in their skill before the middle of the nineteenth

century, and their colleges at Woolwich, Chatham and Addiscombe (the latter was primarily an East India Company Institution) deserve recognition for their pioneering role
in British engineering education. Nevertheless, for all their distinctions these prominent
military engineers made little direct contribution to British domestic (as distinct from colonial) engineering, except through certain forms of specialist oversight such as that of
the railway inspectorate for which they were well suited by virtue of having no financial
involvement in the enterprises subject to their review. Governments tended to turn frequently to the military engineers for advice, but their direct contribution to engineering
within Britain was minimal.

The main contribution of engineering expertise to the industrial and commercial activity of the nation came thus from the other two thirds of the engineers listed in the *DNB*,
and particularly from the 129 who can be identified as fully professional engineers. These
were the men who were setting the pace of development in industrial innovation and in
transport and communications. They performed the pivotal role of the professional engineers in the transformation of British society in the period with which this study is concerned. The *DNB* selection of men in this vital category serves both to emphasise the high
regard of the Victorian public for the leading professional engineers, and to show how
the services of a host of other engineers had come to be taken for granted as a matter of
routine.

Another biographical exercise, coinciding with the first edition of the *DNB* at the turn
of the nineteenth century, was that conducted by Frederic Boase, whose *Modern English
Biography* was published in three volumes between 1892 and 1901, with three supplementary volumes following between 1901 and 1921.[42] The compilation is rather different from
that of the *DNB*, with a focus upon people who died between 1851 and 1900, and a short
factual report on each entry. But it remains a remarkable fact that two such similar publications should have appeared about the same time, and it serves to demonstrate the development of prosopographical investigations in the period with which we are concerned.[43]

Unlike the *DNB*, Boase gives a full index to each volume, including an entry for 'Engineers'. It is thus possible to make an easy calculation, whereby it can be claimed that
out of a total of some 30,000 short biographical sketches of persons who died in the second half of the nineteenth century, 198 are engineers. This is a proportion of one in 150
compared with the one in sixty-two for the *DNB* out of a similar total of entries. But
against this comparison it should be remembered that the categories included in the *DNB*
analysis are more widely cast than those in Boase, and that many who were classified as
military and general engineers in our discussion of the *DNB* could have been included
under headings other than 'engineer' in Boase. The Boase figure of 198 engineers is best
compared with the 129 in the *DNB* category of 'professional' engineers, and in this light
it can be seen as slightly more generous towards late nineteenth century engineers than
the *DNB*. Certainly the compilers of Boase seem to have had second thoughts about many
of these engineers, because the number was almost doubled (from 104 to 198) in the
three supplementary volumes, and included several substantial engineers who did not
reach entries in the *DNB* such as Joseph Mitchell (1803-1883), William Barber Buddi-

com (1816-1887), John Galloway (1804-1894), George Fosbery Lister (1822-1899), Peter William Willans (1851-1892), and Samuel Hansard Yockney (1813-1893).

For the most part, however, the resemblances between the selection of engineers for entries in Boase and the *DNB* are more remarkable than any differences of emphasis, and the cross-section of the engineering community displayed is essentially the same. That is, there is a preponderance of civil engineers, with railway works figuring prominently, together with gas, water and harbour works, with mining and overseas engineering receiving a good coverage. Naturally enough, considering the cut-off date of 1900, few electrical engineers appear, and there is little indication of the institutional proliferation of the late nineteenth century. For example, John Ramsbottom, pioneer of the Institution of Mechanical Engineers, is not mentioned in either Boase or the *DNB*. The same applies to Thomas Newbigging, early president of the Gas Engineers; Lewis Angell, founder of the Municipal Engineers; Nicholas Wood, moving spirit of the North of England Institute of Mining Engineers; or of Asplan Bedlam, founder of the Marine Engineers. These omissions give some indication of the sub-structure of the engineering community which went virtually unrecognised in the non-engineering world. The evidence of the national biographical exercises thus confirms the tendency to polarise the engineering profession between a comparatively small group who became public figures, and a largely anonymous mass of members about whom it is difficult to reconstruct any precise profiles of their careers and personalities.

Detailed analysis of the biographical evidence available through obituaries might eventually make possible a more general prosopographical reconstruction of the nature and quality of the British engineering profession in the nineteenth century than that available from the standard biographical compilations, but it is unlikely that the profound anonymity shrouding nine-tenths of the members of the professional institutions will be dispersed sufficiently to reveal anything like a complete picture. It is possible, however, to make a much closer study of attitudes of the engineers to education and training, and to their social, political and religious affiliations, than anything that has been attempted hitherto. The prosopographical method has been convincingly expounded by Professor Lawrence Stone and applied to good effect, for example, in the analysis of British parliamentarians. Steven Shapin and Arnold Thackray have recommended its application to the British scientific community in the eighteenth and nineteenth centuries.[44] It is intended to make use of such insights in the present study, but for the time being we must continue to rely mainly on such generalisations as can be made from the lives of the better known engineers. This can take us quite a long way, as will become apparent in the following pages.

Even though the question regarding the quality of the majority of British engineers in the nineteenth century is thus only capable at present of a partial and impressionistic answer, biographical material by no means exhausts our sources of engineering history. It has already been suggested that one reason for the decline in engineering biography after 1860 could be the switch to greater emphasis on institutional organisation which occurred from the middle of the nineteenth century. If this is so, it must be added that it is not reflected in the quality of such institutional histories as are available, because most of these make exceedingly dull reading. But what does become available, from the 1840s onwards, is a large body of material produced by the institutions, and especially the impressive series of volumes of *Proceedings* and *Transactions* beginning with those published by the Institution of Civil Engineers.[45] Much of this substantial body of literature is understandably devoted to the discussion of highly technical matters germane to the field of interest to the institution concerned, but it also contains information about the business of the institutions, including comments on size and composition of their membership, disciplinary matters, and current issues of general concern to members. The *Transactions* of the institutions usually contain regular presidential addresses in which the chief officer of the society takes the opportunity presented by his inauguration into office to give a general survey of the affairs of the profession, and even though these tend to be repetitive and unadventurous in content they occasionally contain passages of considerable illumination. These, together with the regular engineering journals which appeared in the second half of the nineteenth century - *The Engineer* was first published in 1856 and *Engineering* in 1866 - provide a rich fund of documentary material from which the development of the institutional history of the engineers can be reconstructed.

No such general history of the institutional development of British engineering in its social context has hitherto been written. The subject lay beyond the didactic intention of Samuel Smiles, so that it only received oblique and incidental treatment in his work. L. T. C. Rolt was concerned primarily with engineering biography, although he did apply himself very successfully to more general accounts of engineering history, and to the history of the Institution of Mechanical Engineers.[46] Unlike Smiles, Rolt was himself an engineer by training and inclination, so that he was able to understand engineering technicalities which he could then expound in lucid prose for the benefit of the lay reader. Few other noteworthy scholars have applied themselves to this subject. The institutional histories which have been published are largely uninspired chronologies, with no attempt to place their subjects in a meaningful historical context. The *Transactions of the New-comen Society* have contained valuable articles on the early Society of Civil Engineers, but no comparable study of any subsequent institutions.[47] The most significant general study has been that provided by W. H. G. Armytage in his pioneering work of 1961, but this serves only to outline the scope of the problems requiring full documentary investigation.[48] There is thus ample scope for such an investigation, which will be attempted in the following pages. It must be recognised, however, that a complete institutional history is beyond the scope of the present study which, while drawing on the histories of many institutions, only pursues them in relation to a general theme of institutional development. No attempt is made to deal in detail with any one institution.

This introductory chapter has been concerned with establishing the scope of engineering history and with basic definitional and methodological aspects of the following study. It is intended now to pursue the main themes of engineering development, seeking first to present a basically chronological account of the emergence and proliferation of professional institutions from the mid-eighteenth century through to 1914, then to examine the significance of these developments in terms of the training of engineers, and the spread of engineering to other parts of the industrialising world. This is not a technical history, so that there will be no attempt to make a technical evaluation of any engineering structures or machines. Doubtless it helps to understand engineering achievements to have some knowledge of the physics of structures; of the technical problems of fluid dynamics, metal fatigue, corrosion, soil analysis, and so on; and even of the laws of thermodynamics. For our purposes, however, in attempting to make an historical assessment of the development of the engineering profession, such technical expertise is not strictly relevant and there will be no discussion of it, even though we will be constantly sensitive to the physical evidence of the subject in the shape of the impressive monuments of nineteenth century engineering in Britain. After all, these are the artifacts which engineering history is all about.[49]

It is hardly necessary to justify the limitation of this examination of engineering professionalisation to Britain, because the process began in Britain as past of the accelerating industrialisation in which this country led the rest of the world, and it followed a pattern of development which was, at least in its early stages, remarkably independent of foreign initiatives, so that it is convenient to treat it separately. The fact that British engineers eventually exerted a profound influence on developments elsewhere is taken into account, but the basis of the treatment is emphatically that implied in taking a distinctly British point of view. Similarly, our study starts in the middle of the eighteenth century for the obvious reason that it was at that time that the first signs of engineering professionalism emerged. It ends in 1914 for reasons which are less obvious, but are nevertheless strong. The twentieth century brought important new departures in engineering, especially with the widespread adoption of electricity and the internal combustion engine. Moreover, the First World War marked a significant break in the development of the engineering institutions, introducing a move towards reintegration rather than continued proliferation. The outbreak of the war in 1914 provides a convenient point, therefore, at which to conclude an account of the dominant trends in nineteenth century British engineering. Subsequent events are a somewhat different story, and require another sort of treatment.

Notes

1. Marxist analysis, based on the differing relationships of conflicting social classes to the means of production, made no specific provision for the role of professional groups. See K. Marx, *Capital*, 1867, Penguin ed. 1976, where in 940 pages of text covering the evolution of modern industrial society there is no discussion of professionalism as a class phenomenon.

2. Harold Perkin, *The Origins of Modern English Society 1780-1880*, Routledge, London, 1969, pp. 256 and 258. The passage continues: '...they had a professional interest in disinterestedness and intelligence. It was their interest to "deliver the goods" which they purveyed: expert service and the objective solution of society's problems, whether disease, legislation, administration, material construction, the nature of matter, social misery, education, or social, economic and political theory' (p. 260).

3. Quoted Perkin, *op. cit.* p. 266.

4. J. Burnham, *The Managerial Revolution*. Penguin, London, 1941.

5. A. Carr-Saunders and P. A. Wilson, *The Professions*, Oxford. 1933, p. 287. The *Shorter Oxford English Dictionary* omits the words: 'or science'.

6. R. Lewis and A. Maude, *Professional People*, Phoenix House, London, 1952, p. 54.

7. Ibid, p. 54: see also p. 74, note 2.

8. Ibid, pp. 55-56.

9..Ibid, p. 57, footnote *

10. W. J. Reader, *Professional Men: the rise of the professional classes in nineteenth century England*, Weidenfeld, London, 1966, p. 145.

11. Quoted Reader, *op. cit.* p. 1.

12. Francis Bacon, *The Elements of the Common Laws of England*, 1630. As a recent example of a use of these words in support of professional aspirations, see the title of the book by Edwin Green, *Debtors to their Profession - A History of the Institute of Bankers 1879-1979*, Methuen, London, 1979.

13. Barrington Kaye, *The development of the architectural profession in Britain*, London, 1960.

14. F. M. L. Thompson, *Chartered Surveyors: the growth of a profession*, Routledge, London, 1968.

15. E. Green, *op. cit.*

16. See R. A. Buchanan, 'Gentlemen Engineers: the making of a profession', *Victorian Studies*, 26, . 4, Summer 1983, 407-429. Also Edward Hughes, 'The professions in the eighteenth century', *Durham University Journal*, new series, 13, 1952, pp. 46-55 - 'the high standards and traditions of professional conduct ... derive from the eighteenth century conception of gentility' (p. 55).

17. Carr-Saunders and Wilson, *op. cit.* pp. 155-165.

18. Reader, *op. cit.* pp. 70-1

19. Martin J. Wiener, *English Culture and the Decline of the Industrial Spirit 1850-1880*, Cambridge, 1981.

20. J. E. Gerstl and S. P. Hutton, *Engineers: the anatomy of a profession - a study of mechanical engineers in Britain*, Tavistock Press, London, 1966.

21. Baroness Wootton observed many years ago that it is one of the few principles of the British wage and salary structure that leaders are always paid more than those they lead: see Barbara Wootton, *The Social Foundations of Wage Policy*, Allen and Unwin, London, 1955, p. 67.

22. See J. B. Jefferys, *The Story of the Engineers. 1800-1945*, Lawrence and Wishart, London, 1945, p. 37, which is concerned with the early development of the craft union of artisan engineers. The identification of Thomas Fairbairn as 'Amicus' is confirmed in W. Pole (Ed.), *Life of Sir William Fairbairn, Bart*, Longman, London, 1877 (reprinted by David and Charles, Newton Abbot, 1970), pp. 323-7. The prejudice against trade unions was widely shared amongst professional engineers at this time.

23. S. Smiles, *Autobiography* (ed. Thomas Mackay), John Murray, London, 1905. See also A. Briggs, *Victorian People*, Odhams Press, London, 1954, where chapter V, pp. 125-149 gives a useful summary of Smiles' career. Briggs quotes Gladstone as agreeing with Smiles that: 'the character of our engineers is a most signal and marked expression of British character, and their acts a great pioneer of British history' - p. 131.

24. Smiles' main works of engineering history were: *Life of George Stephenson* (1857); *Lives of the Engineers* (3 volumes, 1862): David and Charles edition, Newton Abbot, 1968, with introduction by L. T. C. Rolt; *Lives of Boulton and Watt* (1865), *Men of Invention and Industry* (1884); and *James Nasmyth* (1885).

25. W. Pole, see note 22 above.

26. O. J. Vignoles, *Life of Charles Blacker Vignoles*, 1889. A more recent study, also by a member of the family, is K. H. Vignoles, *Charles Blacker Vignoles, Romantic Engineer*, Cambridge, 1982.

27. T. Mackay, *The Life of Sir John Fowler, Engineer, Bart.*, John Murray, London, 1900.

28. H. W. Dickinson and Rhys Jenkins, *James Watt and the Steam Engine*, 1927, republished Moorland, Ashbourne, Derbyshire, 1981: H. W. Dickinson, *James Watt, Craftsman and Engineer*, Cambridge, 1935; H. W. Dickinson, *Matthew Boulton*, Cambridge, 1936; H. W. Dickinson and Arthur Titley, *Richard Trevithick - the Engineer and the Man*, Cambridge, 1934.

29. Sir Alexander Gibb, *The Story of Telford - The Rise of Civil Engineering*, Maclehose, London, 1935; R. Beamish, *Memoir of the Life of Sir Marc Isambard Brunel*, Longman, London, 1862; Lady Celia Brunel Noble, *The Brunels - Father and Son*, Cobden-Sanderson, London, 1938; I. Brunel, *The Life of Isambard Kingdom Brunel, Civil Engineer*, Longman, London, 1870, republished David and Charles, Newton Abbot, 1971.

30. Joseph Mitchell, *Reminiscences of my Life in the Highlands*, published privately in 2 volumes, 1883, republished in 2 volumes, David and Charles, Newton Abbot, 1971: For description of the Civils, see volume 1, pp. 87-100.

31. *John Brunton's Book, Being the Memories of John Brunton, Engineer...*, with introduction by J. H. Clapham, Cambridge, 1939.

32. Sir Daniel Gooch, *Memoirs and Diary*, edited by Roger Burdett Wilson, David and Charles, Newton Abbot, 1972.

33. See particularly L. T. C. Rolt, *Thomas Newcomen - The prehistory of the steam engine*, David and Charles, Dawlish, 1963: republished as L. T. C. Rolt and J. S. Allen, *The steam engine of Thomas Newcomen*, Moorland, Hartington, 1977.

34. Stevenson's family had complained about denying him credit for the Bell Rock Lighthouse: see D. Stevenson, *Life of Robert Stevenson, civil engineer*, Edinburgh, 1898. For a modern account, see C. Mair, *A Star for Seamen - the Stevenson Family of Engineers*, John Murray, London, 1978.

35. The Footnote appears in the course of 'The Life of John Rennie' on pp. 197-8 of volume 2 of the 1968 edition of *Lives of the Engineers*. See also C. Hadfield and A. W. Skempton, *William Jessop, Engineer*, David and Charles, Newton Abbot, 1979.

36. George S. Emmerson, *John Scott Russell, A Great Victorian Engineer and Naval Architect*, John Murray, London, 1977. I have argued elsewhere that Emmerson overstates his case by attacking Rolt, but this does not invalidate his main point even though he fails to present Scott Russell as an attractive character: See R. A. Buchanan, 'The *Great Eastern* Controversy: A Comment', in *Technology and Culture*, 24, January 1983, 98-106.

37. See the discussion below, chapter five, on the growth of numbers of professional engineers. The position is complicated by the practice of dual or multiple membership of institutions, but the figures given here provide a sound basis for generalisation.

38. S. P. Bell, *A Biographical Index of British Engineers in the Nineteenth Century*, Garland, New York and London, 1975.

39. See Mackay, *Fowler*, p.274 and p.311.

40. The original proprietor of the *DNB* was George Smith and much of the editorial work was done by Sir Sydney Lee. It was published in 1903 by Smith, Elder and Co., although subsequent printings were made by Oxford University Press. It covered the period from the beginning of British history to 1900 in sixty-three alphabetical volumes, and three supplementary volumes were added later.

41. For further details on James Douglas see Barry M. Marsden, *Pioneers of Prehistory - Leaders and landmarks in English archaeology (1500-1900)*, G. W and A. Hesketh, Ormskirk and Northridge, 1984, p.12

42. Boase was re-published in six volumes by Frank Cass in 1965.

43. The Editor of Boase remarked in the Preface to the first edition: 'Biography like other subjects seems to have its fashion; at one time it is much attended to, at another time neglected'.

44. Lawrence Stone, 'Prosopography', in *Daedalus*, Winter 1971, 46-79; L.B. Namier, *The structure of politics at the accession of George III*, London, 1929; and Steven Shapin and Arnold Thackray, 'Prosopography as a research tool in the history of science: the British scientific community 1700-1900', in *History of Science*, 12, 1974, 1-28.

45. The *Minutes of Proceedings* of the Institution of Civil Engineers were first published in 1841 and the series has continued ever since: three volumes of *Transactions* appeared separately in 1836, 1838, and 1842, but thereafter they were incorporated in *Proceedings*: see below, Chapter 4.

46. See, for instance, L. T. C. Rolt, *Victorian Engineering*, Allen Lane, Harmondsworth, 1970; and *The Mechanicals - Progress of a Profession*, Heinemann, London, 1967.

47. See, for instance, S. B. Donkin, 'The Society of Civil Engineers (Smeatonians)', in *Trans. Newcomen Soc.*, 17, 1936-7, 51-71; Esther Clark Wright, 'The Early Smeatonians', in *Trans. Newcomen Soc.*, 18, 1937-8. 101-110; and A. W. Skempton and Esther Clark Wright, 'Early members of the Smeatonian Society of Civil Engineers', in *Trans. Newcomen Soc.*, 44, 1967-8, 23-47.

48. W. H. G. Armytage, *The Social History of Engineering*, Faber, London, 1961.

49. The study of the physical evidence of engineering history has become popular in recent decades under the name of 'industrial archaeology'; see R. A. Buchanan, *Industrial Archaeology in Britain*, Pelican, Harmondsworth, 1972.

Chapter Two

The Origins of the Profession

The British engineering profession emerged in the second half of the eighteenth century. It was, in the most general sense, a response to the increasing pace of industrialisation with its demand for more buildings, roads, harbours, and canals. But it was conditioned, in the particular form which it took, by the activities of the precursors of the modern professional engineers, and even more especially by the character of the men who made the formative decisions in the 1750s and thereafter. To understand the origins of British engineering as a self-conscious professional group, therefore, it is necessary to examine the relationship between these general and particular features, reviewing first how the condition of industrialisation occurred, and then placing in this context the engineering personalities who experienced it and who were able to grasp the opportunities which it offered. On the basis of such an analysis it should then be possible to provide an interpretation of the first formal association of engineers for professional purposes - the Society of Civil Engineers, founded in 1771.

It has become fashionable amongst British scholars to regret the term 'Industrial Revolution' on the grounds that it arouses unjustified expectations of a sharp discontinuity in historical development whereby an agricultural - or at least pre-industrial - society was quite suddenly transformed into a land of textile mills and iron forges, driven by steam engines the furnaces of which filled the air with the smoke of burning coal through a forest of chimneys. The fact that such a transformation did occur in Britain is not seriously challenged: the point at issue is the speed of the change, and the period of time over which it took place. Early twentieth century text books spoke confidently of the change beginning about 1760 and being virtually complete by 1830.[1] Subsequently, historians have become less sure about these dates. They have recognised a long period of preparation for the rapid changes in certain sectors of industry, going back to the sixteenth and seventeenth centuries, and even having anticipations in the Middle Ages.[2] And they have come to see that, far from finishing in or near 1830, the process of industrialisation continued to gather momentum, spreading from the original industrial leading sectors to all other aspects of industrial and social activity, and being conveyed to other countries which then undertook their own process of rapid industrialisation.[3]

The Industrial Revolution has thus come to be regarded as a long-term and continuing process rather than as a sudden transition. Nevertheless, the evidence of a significant quickening of the process in eighteenth century Britain is overwhelming. Equipped with Newcomen-type atmospheric steam engines, after the first successful installation of such a pumping machine near Dudley Castle in 1712, the coal mining industry in the Midlands and North East expanded steadily, while the steam engine was applied to other water-pumping functions and spread to continental Europe and North America.[4] Equipped with Abraham Darby's process for producing cast-iron from coke-fired blast furnaces, first established at Coalbrookdale in 1709, the British iron and steel industry made substantial advances in its performance. British pig iron production increased from 17,350 tons in 1740 to 678,417 tons in 1830.[5]

With the aid of Cort's puddling process, introduced in 1784, the manufacture of wrought iron underwent a corresponding expansion, while Huntsman's high-quality homogeneous crucible steel was becoming widely available by the end of the century. And equipped with the novel textile machines which followed John Kay's 'flying shuttle' - Samuel Crompton's 'spinning mule' and Edmund Cartwright's 'power loom' amongst many others - the British textile industry, and especially the manufacture of cotton fabrics, was converted between 1750 and 1830 from a comparatively small 'domestic' industry into a large-scale factory industry concentrated strongly in the mill towns of Lancashire, Derbyshire, and Yorkshire.[6]

Stimulated by these developments, the total volume of trade increased rapidly and created a pressing demand both for better port facilities to handle the growth of imports and for better inland transport to make easier the movement of goods. In response to the first need, the major ports all undertook the construction of complex systems of enclosed docks offering specialised trading facilities, and a determined attempt was made to improve port works and lighthouses.[7] In response to the second need, various measures of road improvement were promoted, and from 1760 onwards for a whole generation British entrepreneurs and financiers expended tremendous resources on the construction of a network of canals.[8] Meanwhile, landowners all over the country had been seeking ways of maximising the profitability of their estates by enclosure and consolidation into large farms, by land drainage and soil improvement, and by the introduction of new crops and better animal stock.[9] Food production increased correspondingly, and was able to meet the basic needs of a rising urban population, while the manufacture of beer and strong spirits, of furniture and clothing, glassware and ceramics, and a host of other commodities, all prospered. Even though there was no dramatic revolution involved in these developments, they profoundly modified the quality of British life.

Just as the consensus of scholarly opinion accepts the fact of an acceleration in the tempo of industrialisation in the eighteenth century, so it is agreed that this process occurred first in Britain. Indeed, it is demonstrable that it occurred *only* in Britain during the eighteenth century. France was the wealthiest and most populous country in Europe at the beginning of that century, and it would have been reasonable to expect new industrial developments to occur there. In some respects, such as in scientific and educational developments, French leadership in Europe remained unchallenged throughout the century. But exhaustion from the wars of Louis XIV and the stultifying effects of the *ancien regime*, even before the severe disruption of the Revolution in 1789 and the subsequent Napoleonic Wars, prevented the industrial realisation of the potential for economic growth in French society until well into the nineteenth century. With the Netherlands in decline after their seventeenth century Golden Age, and Sweden in comparative isolation after its defeat at the hands of Peter the Great's Russia; with Spain lapsing into a third-rate power and with the states of Germany hardly yet recovered from the destruc-tive effects of the seventeenth century Thirty Years War, there was no other European rival to challenge the steady rise of Great Britain to a world domination based on exten-sive trading relationships, an enterprising colonial system, and rapid industrialisation.[10]

Britain was able to take advantage of this international situation which favoured its growth in the eighteenth century because of a fortunate combination of internal factors. For one thing, it had emerged as a firmly united kingdom, consolidated by the Act of Union between England and Scotland in 1707. It was also a kingdom enjoying the bene-fits of constitutional monarchy whereby the executive and legislative machinery was based on a system of parliamentary representation in which the landed and propertied classes were able to assert their interests. This had been achieved through the seven-teenth century conflict between monarchy and parliament, consummated in the settlement of 1688 - the so-called 'Glorious Revolution' in which the absolute power of the monarchy was finally destroyed without bloodshed. As a result of these developments, Britain had come to possess a flexibility and a readiness to adopt administrative and fin-ancial innovations which was not shared by its great rival across the Channel. Also unlike France, Great Britain was a Protestant power, but one wherein a degree of religious toler-ation had been established in the settlement of 1688, which enabled Huguenots and other religious refugees from continental despots to find a congenial home in which to prac-tise their industrial skills. And almost by accident, Britain had acquired a string of flour-ishing colonies on the eastern seaboard of North America which provided a valuable basis for trade and for mercantile and naval strength.

The quickening of industrialisation in Britain in the eighteenth century was thus so-lidly based. For the previous two centuries, under the Tudors and Stuarts, the sinews of industrial growth had been nourished by governments pursuing a policy of mercantilist control in order to ensure the strength and independence of the state in any conflict. There had even been some preliminary experience of the sort of large-scale operations which were subsequently to play such an important part in promoting the rise of the en-gineering profession. For instance, German mining engineers had been encouraged to come to England to apply the skills already made famous by the publication of Georgius Agricola's *De Re Metallica* in 1556 to winning metals in the Lake District and elsewhere;[11] Hugh Myddelton had implemented an ingenious scheme to improve the London water supply by his 'New River', completed in 1613, and won the first engineering accolade for this and other services;[12] and several Dutch engineers were persuaded to bring their skills in land reclamation to aid English adventurers in improving their estates. The outstand-ing representative of the latter enterprise was Cornelius Vermuyden, born in St. Maur-tensdijk, Zeeland, in 1590, who came to England in 1621 at the instigation of Dutch colleagues who were negotiating terms with a party of English landowners hoping to undertake the drainage of the Fenland. This party was led by the Russell family, Earls and later Dukes of Bedford, who had managed to win the support of King James I for the project. Nothing came of the plans immediately, but Vermuyden received other drain-age commissions, and his success in these ventures, although only partial, led to him re-ceiving a knighthood from King Charles I in 1629.[13]

By 1630, Vermuyden had been engaged by Francis, the fourth Earl of Bedford, as engineer for the Great Level of the Fens, and his greatest work could at last commence. Despite difficulties with the proprietors, local residents, shortage of capital, and the many problems caused by the political disruption of the Civil Wars, he completed this task in 1655. The Great Level constitutes only a third of the total area of the Fenland, but Vermuyden's methods of controlling the flow of the Great Ouse were to set the pattern for all subsequent land drainage schemes in the area. These methods did not go without challenge: Vermuyden's compatriot Westerdyke advocated deepening and banking existing river courses, whereas Vermuyden favoured the cutting of new straight courses, with banks set well back from the edges of the waterway to allow for 'washes' or artificial flood plains. There was plenty of scope for both methods in the Fens, but Vermuyden's solution made the larger contribution to the landscape, with the twin 'Bedford Rivers' in particular running in parallel straight lines across the flat expanse of the Great Level for twenty one miles, and providing a substantial monument to the engineer who designed them.[14]

Sir Cornelius Vermuyden appears to have been a disputatious and obstinate man, to judge from the long list of controversies and litigation with which his career was punctuated. His interests were not confined to land drainage: he owned property in several parts of the kingdom, including King's Sedgemoor in Somerset, which he acquired from Charles I but to which he made no significant drainage contribution, and he held a share in a lead-mining partnership at Wirksworth in Derbyshire. Despite his recognition by the early Stuart kings, he seems to have established a good relationship with Oliver Cromwell, who used his services in delicate negotiations for a treaty with the Dutch in 1653, and with the collapse of the Protectorate Vermuyden's career seems effectively to have terminated. He died in comparative obscurity in Westminster in 1677, having presumably lived off his various estates for the last twenty years of his life. It is difficult to form a clear conception of Vermuyden's sense of his role as an engineer. His modern biographer speaks of his consciousness of 'professional integrity'[15] but it is unlikely that he saw himself as in any way belonging to a community of engineers, and in so far as he did it would presumably have been in relation to his Dutch colleagues. He had nothing to learn in the way of engineering expertise from his English contemporaries, and his main service - apart from the very practical gain in property values which his schemes achieved - was that of pointing the way ahead for a later generation of land drainage engineers, who were able to approach the same problems which he had tackled, but with greater resources of capital and mechanical aids. In this sense he is a genuine precursor to subsequent engineering developments.

The knighthoods granted to Sir Hugh Myddelton and Sir Cornelius Vermuyden in the 1620s may be seen symbolically as a national recognition of the importance of engineers, but the fact that there were no further honours of this kind for two hundred years does not mean that engineering became less important. Far from it: the tempo of engineering works increased steadily in the eighteenth century, and the lack of formal honours in this century reflected a changing view of such political rewards by both politicians and engineers. Nevertheless, the British engineers in the first half of the century remained a widely scattered group of individuals with hardly any sense of collective identity other than that derived from sources outside the still embryonic profession. We may review these sources under the categories of sectarian, craft, dynastic, and military relationships.

The best illustration of sectarian associations occurs early in the century, amongst the first generation of steam engine builders. Despite the great significance that the introduction of steam power was later to have for the development of the engineering profession, these pioneers do not seem to have thought of themselves as 'engineers'. Thomas Newcomen himself was 'a practical tradesman, an ironmonger of Dartmouth', and to the end of his career he was 'proud to describe himself as "Ironmonger"'.[16] His partner, John Calley, has been described as 'a plumber and glazier' (though the 'plumber' could be a later addition).[17] The Potters and Hornblowers, who were amongst the first people to act for the Proprietors of the Newcomen engine and formed dynasties of engine builders responsible for transferring Newcomen engine technology to continental Europe and America, do not appear to have had any engineering background before their involvement with Newcomen.[18] Of far more importance than any professional relationships between these early steam engineers was their common religion, as they were nearly all nonconformists of Baptist persuasion. While the people called Quakers established an early dominance in the British brass and ironfounding industries, with the activities of such families as the Darbys and Champions, steam technology seems to have been appropriated by the Baptists. The economic significance of these relationships was that they made available a network of contacts, often related by marriage, of people who could be trusted as agents and as sound investments for surplus capital. But the social functions of the religious links should also be acknowledged, because they provided a substitute for professional associations and, indeed, made the latter superfluous.[19]

Of course, knowledge of the steam engine and demand for its services quickly moved beyond the scope of a tiny religious sect. Hard-headed businessmen like Stonier Parrot and George Sparrow, foreign adventurers like Marten Triewald, and scientific popularisers like J. T. Desaguliers, all applied themselves in various ways to exploiting the possibilities of the new prime mover.[20] Even more to the point, it seems that after the lapse of the controlling Savery patent in 1733 millwrights and other craftsmen all over the country were prepared to lend their hands to the construction of steam engines as required by their clients. A not untypical example is that of the millwright James Brindley, before he embarked on the career as a canal engineer which gave a new dimension to his reputation. Brindley was born in 1716. His father acquired a farm near Leek in Staffordshire, and apprenticed his son to a millwright in Macclesfield. James appears to have set up in business as a millwright in 1742, and he moved to Burslem, to be near the industrial centre of the Potteries, in 1750. His work in the 1750s was very varied, and included the construction of several steam engines, apparently introducing his own modifications to the basic Newcomen design. But he also undertook to construct watermills and river-control works before his talents were recognised by the Duke of Bridgewater who employed him to build the first stretch of the Bridgewater Canal in 1759 and thus initiated the translation of Brindley from millwright to canal engineer.[21]

The skills which Brindley had learnt as a millwright were a mixed bag of odd-jobbing, carpentry, metal-working, and basic surveying. Years later, they were vividly described by Sir William Fairbairn, reviewing his own long career in which he had moved from being a travelling millwright to become one of the most distinguished engineers of his generation. In an often-quoted passage he described the skills which he, like other eighteenth century millwrights, had acquired in the course of a long apprenticeship and practical experience:

> '... the millwright of the last century was an itinerant engineer and mechanic of high reputation. He could handle the axe, the hammer, and the plane with equal skill and precision; he could turn, bore or forge with the ease and despatch of one brought up to these trades: and he could set out and cut in the furrows of a millstone with an accuracy equal or superior to that of the miller himself. These various duties he was called upon to exercise, and seldom in vain, as in the practice of his profession he had mainly to depend upon his own resources. Generally, he was a fair arithmetician, knew something of geometry, levelling, and mensuration, and in some cases possessed a very competent knowledge of practical mathematics. He could calculate the velocities, strength, and power of machines, could draw in plan and section, and could construct buildings, conduits, or water-courses, in all the forms and under all the conditions required in his professional practice: he could build bridges, cut canals, and perform a variety of work now done by civil engineers.'[22]

These skills resembled those of such crafts as stone-masonry, which provided other sources of craft-trained engineers, with the career of Thomas Telford one of their outstanding success-stories. They were skills which were at a premium in a period of rapid industrialisation, with the need for new construction works of many kinds and of an ever-increasing magnitude, so that young men of talent who had acquired them were, in some cases, able to make the transition into the new profession of civil engineering. Of course, only a small minority possessed both the ability and had the opportunity to make the transition, but those who did came to provide one of the most important sources of recruitment to the profession, and the emphasis on learning by practical experience remained as one of the distinguishing characteristics of British engineering thereafter.

Another significant sub-professional group which became important in the development of British engineering was dynastic. The history of engineering has witnessed the success of many such dynasties. One of the most interesting examples in the eighteenth century was the Hornblower family, which achieved a remarkable continuity of engineering talent over three generations, and the subsequent importance of family connections amongst the Rennies, the Stephensons, the Stevensons, and the Brunels, is self-evident.[23] This is not to say that the younger generations of these families were able to become engineers with less training than less fortunate aspirants to the profession: on the contrary, their fathers were usually able to ensure that they received a particularly thorough and relevant education. But there can be little doubt that being the son or grandson of an established engineer was a considerable advantage in giving access to senior posts at an earlier age than would otherwise have been the case. Sometimes young engineers who were not directly related to members of the profession were able to benefit from a surrogate dynastic relationship by virtue of having served a specially intimate apprenticeship. Thus the young William Jessop, the son of John Smeaton's site supervisor on the Eddystone Rock Lighthouse, was brought up in Smeaton's home as a pupil,[24] and we have it on the authority of Joseph Mitchell, canal and railway builder in the Highlands of Scotland, that Thomas Telford regarded all his pupils as his sons.[25] Similarly, the first John Rennie acquired proficiency at the bench in the workshop of Andrew Meikle, and

Henry Maudslay trained a whole generation of mechanical engineers who went on to become leading members of the profession.[26] Such dynastic networks become less important with the growth of the professional institutions, but they made a formative contribution to the early stages of British engineering.

In addition to sectarian, craft, and dynastic relationships, another source of engineering skills in the eighteenth century was the Army. As the term 'civil engineering' seems to have arisen to distinguish its practitioners from the more familiar military engineers, it would be reasonable to expect there to be a substantial military contribution to the birth of the profession. Any such expectation, however, is disappointed by the available evidence. It is true that the tradition of military engineering in Britain is a long one, going back at least to the 'Waldivus Ingeniator' who was Chief Engineer to King William I. But for several centuries the role of such engineers was shadowy and episodic. By the time of the French Wars at the end of the seventeenth century, experience of the great military fortresses established on the continent by Vauban and other military engineers had encouraged the practice of equipping 'trains' of artillery, military supplies, and engineers, to accompany the army with a 'Chief Engineer', 'Sub-Engineers', and other staff. A number of these trains were commissioned for service on the continent under Marlborough, the role of the engineers being usually that of preparing fortifications or undermining those of the enemy (hence, they became known as 'sappers'). After the wars, in 1716, there was a significant reorganisation as a result of which the Royal Engineers became a distinct Corps while the artillery was separated from it to become the Royal Regiment of Artillery. A further warrant of 1717 created separate establishments for Minorca and Gibraltar, the defence of which became a primary commitment of the new Corps.[27]

During much of the eighteenth century, the traditional British dislike of a 'standing army' led to the curtailment of all military activity, so that there was little scope for the development of the Corps of Royal Engineers, except for a limited function in surveying and mapping. Officers of the Corps - Major General David Watson and later Major General Roy - were made responsible for the official survey of the Scottish Highlands beginning in 1747, and this became an important prelude to the Ordnance Survey of the whole country.[28] The renewal of a continental commitment for the British Army in the Napoleonic Wars at the beginning of the nineteenth century gave a further opportunity for the development of the Royal Engineers. In particular, they made their mark in the Peninsular War in Spain and Portugal, the lines of Torres Vedras being almost entirely their creation. It was experience in this campaign which convinced Captain Charles Pasley R.E. 'how deficient in the proper knowledge of field engineering were not only the men of the Royal Military Artificers, but also the officers of the Corps', so that he mounted a vigorous argument for the establishment of an engineering school.[29] This was set up at Chatham in 1812 with Pasley as the first Director, and in subsequent decades this institution went on to become a powerful instrument in determining the development of British military engineering. The fact remains, however, that far from giving a lead to civil engineering in Britain, military engineering lagged behind it in most respects in the eighteenth century.

One outstanding instance of military involvement in engineering in the eighteenth century - although not by the Royal Engineers - was the construction of roads in the Highlands of Scotland. These were begun in 1725, when Major-General George Wade (1693-1748), recently appointed Commander-in-Chief, North Britain, went to Scotland with a commission to impose a new Disarming Act on the Highland clans, and to take other measures necessary to guarantee the preservation of law and order in that remote and unsettled part of the kingdom.[30] These measures included the construction of military roads to facilitate the movement of troops and supplies. Several hundred miles of such roads were built under General Wade and Major William Caulfield, who became Wade's Inspector of Roads in 1733 and continued the work of road building and maintenance from 1740, when Wade was promoted to other offices, to 1767. Wade appears to have been an intelligent and humane leader, able to keep up good morale amongst his men, but there is nothing in his career to indicate any familiarity with engineering, either military or civil, and he does not seem to have relied on anybody with much engineering expertise. It appears that road-building in the eighteenth century was regarded mainly as a matter of applied common sense, and that the only things necessary to undertake it were adequate finance and the supplies required by the labour force. As the finance was an annual government grant inspired by recurrent fear of Jacobite rebellions, and as the labour was provided by bands of troops enabled by Wade to earn double pay on road work, the work was able to proceed regularly throughout the summer months as a military exercise.

Road work began each season with a small party - usually one officer with a handful of NCOs and troops - surveying the alignment with a theodolite and chain. Then the main work-force moved in, digging out the roadway and setting in place boulders, broken stone, and gravel to provide a causeway eighteen feet wide between embankments and ditches, the latter connected by occasional open cross-drains to keep any water off the road surface. The engineering skill, such as it was, was confined largely to the choice of route and general alignment, and it seems as if Wade gave his personal attention to these matters, as he spent considerable time in the field each season for eight years, and came to speak affectionately of his 'highwaymen'. Bridges were few and rudimentary, except for that over the Tay at Aberfeldy, for which Wade called upon the architectural skills of William Adam, and the High Bridge at Spean Bridge. The latter was completed in 1736, and was virtually Wade's last construction on the Scottish military road system. As an example of road improvement, the military roads were a conspicuous success. But as an example of the deployment of engineering skills and professional expertise, they are of remarkably little significance.

There was thus no community of engineers in Britain before 1760. Millwrights and other steam engine builders from Newcomen onwards provided machines with growing confidence for their customers - mainly colliery owners and managers, but also the tin and copper adventurers of Cornwall. Others, like George Sorocold and Thomas Steers, undertook to construct the first enclosed docks in London and Liverpool respectively.[31] But most of these men were mechanics, skilled odd-job men, without the resources, the aspiration, or the desire to see themselves as the custodians of a new profession, even if there had been enough of them together at any one time and place to do so. There can be no doubting the skill and even the genius of some of these men. But even if they occasionally saw themselves as in some sense 'engineers' it is certain that the appellation

had no special professional significance for them, because the lack of any large-scale engineering enterprises precluded the emergence of a genuine civil engineering profession
in Britain at the time.

The situation changed dramatically after 1760. The acceleration in the processes of
industrialisation promoted a spate of canal building, bridge construction, harbour improvements, lighthouse building, and new land drainage and reclamation schemes. It also
promoted the construction of factories and other industrial buildings, and stimulated a
demand for power to drive the new machines. These were conditions which at last provided an environment congenial to the development of an engineering community which
began to acquire professional qualities. The first person to recognise the opportunities
presented by this situation was John Smeaton, to whom is usually attributed the first regular use of the term 'civil engineer' some time in the early 1760s.[32] Skempton and Wright
refer to Thomas Yeoman of Northampton as having been described as 'Surveyor and
Civil Engineer' in the 1763 edition of Mortimer's *Universal Director*, and suggest that this
was 'an extremely early use of the latter term'.[33] It is reasonable to speculate that the
term had already entered circulation before Smeaton appropriated it and endowed it
upon the new profession, but Smeaton has rightly come to be regarded as the Father of
Civil Engineering. He seems to have acquired remarkably early a very clear conception
of the professional role of the engineer as a consultant mediating between a client and a
contractor, and to have acted according to this conception in the years between 1759 and
1783 when he practised as a consulting engineer in England, Scotland, and Ireland from
his home at Austhorpe Lodge near Leeds. By 1759 he was already highly regarded in
scientific circles as a fellow of the Royal Society, to which he had been elected in 1753,
and to which he presented his pioneering paper on wind and water power in 1759, and
he had established his reputation as an engineer with the construction of the Eddystone
Rock lighthouse in 1756-59.[34] The fact that he came from a professional background,
his father having practised as a lawyer in Leeds, probably provided Smeaton with his
model of professional standards for the incipient profession, and it also helps to account
for his success in establishing the status of engineering evidence as 'professional' in a
court of law on a par with medical evidence.[35] Smeaton was in great demand as a consultant on canals, bridges, and harbour works. He also applied himself to mill building
and to steam technology and introduced significant improvements in the otherwise standard Newcomen-type engines which he erected at Long Benton Colliery (a 40hp engine,
in 1773) and at Chasewater Mine (a 70hp engine, in 1775). These engines were both built
just before the Watt innovations in the steam engine became available, but Smeaton
showed cautious conservatism about these although he appears to have done little further work on steam engines once Boulton and Watt machines were on the market.[36]

Smeaton's most significant step towards the creation of an engineering profession,
however, was his initiative in the formation of the Society of Civil Engineers in 1771. The
circumstances of this event were the regular encounters of experts employed by canal
companies and other large-scale enterprises to advise them on engineering matters, the
encounters usually occurring in the course of meetings before various parliamentary committees at which the proto-engineers had to represent their clients in order to secure the
legislation necessary for their enterprises. That is to say, the Society of Civil Engineers

can be attributed directly to the rise of large engineering enterprises requiring skilled advice and particularly calling for the presentation of these skills in the context of administrative and legal debate. The delicacy and complexity of these tasks called for abilities and social graces which the jobbing millwright did not usually possess. To be sure, they could be acquired, and James Brindley appears to have acquitted himself very effectively before parliamentary committees.[37] But Brindley died in 1772, and he was not amongst the founders of the Society, although his principal assistant, Robert Whitworth, was. The nature of Brindley's relationship with Smeaton remains something of an enigma, as they were variously described as 'friends' and as 'rivals'.[38] However, it is tempting to think that Smeaton regarded his great contemporary much as he regarded his own loyal resident engineer John Gwyn, of whom he said in excuse for Gwyn's lack of social graces, 'I never recommend him as a writer or as a speaker'.[39] Whether or not Brindley was invited to join the Society we do not know. But from the outset there was an inclination to emphasise the qualities summed up in the English concept of a 'gentleman' as those which were appropriate to membership of the profession. This was not so much snobbishness as a recognition of the value of 'character', integrity, reliability, and such like, amongst people who were striving to assert their professional self-consciousness in a competitive and enterprising society.

An analysis of the list of members of the Society of Civil Engineers compiled by Skempton and Wright reveals several interesting features about the putative profession.[40] Information is available on eight of the eleven members recruited in 1771. Four of these probably had some sort of professional background: Thomas Yeoman and John Grundy as surveyors, Robert Mylne as an architect, and Smeaton as a lawyer. Only Grundy appears to have received any engineering instruction, having been trained as a land surveyor on Fen drainage works by his father. The background of the other four is less clear, but was probably that of millwrights, as Joseph Nickalls had worked as a millwright under Smeaton on Stratford mill and elsewhere, while John Golborne, Robert Whitworth, and Hugh Henshall, had all served under Brindley on canal works. Of the eight, Mylne came from Scotland, Smeaton from Yorkshire, Yeoman from Northamptonshire, Grundy from Spalding, and Golborne from Chester. Three of the eight - Yeoman, Smeaton and Mylne - were fellows of the Royal Society.[41]

Between 1772 and 1792, the effective membership of the Society increased to about twenty, including some 'instrument-makers and other craftsmen' as well as engineers and surveyors.[42] Of the twenty-four listed as having been elected in this period, only six appear to have had any sort of engineering training - John Smith, William Jessop, Henry Watson, John Rennie, Henry Eastburn, and James Watt - although others presumably served their time as millwrights, such as John and James Cooper, while one - John Holmes - was a trained clock-maker. Major Watson was the first member elected from a military engineering career, having been at Woolwich Academy. Three - Joseph Hodkinson, William Faden, and George Young - were cartographers and land surveyors. Canal work provided the most common expertise, with land drainage a close second. At least two - Thomas Morris and Thomas Dadford - had served under Brindley, and several such as John Pinkerton, Smith, and Dadford, had established themselves as engineering contractors. One, Matthew Boulton, was an industrialist, and another, Samuel Wyatt, an architect. Only one was unequivocally a steam engineer - James Watt - although one other, Samuel Phillips, was described as 'Engine Maker to his Majesty's Board of Works', so that he was presumably familiar with the new technology, as indeed was Rennie, whose first large commission had been for the installation of the Albion Mill machinery under Watt's direction. The Scottish contingent in this selection of twenty-four remained small, consisting of Rennie and Watt, while the Yorkshire group had five members - Smith, Holmes, John Gott, John Longbotham, and Joseph Priestley. Four of the twenty-four were elected FRS - Boulton, Watt, Rennie, and Watson.

The Society experienced some unspecified internal dissensions shortly before Smeaton's death in 1792, and in the following year a committee of four members - Mylne, Jessop, Whitworth, and Rennie - was appointed to reorganise it in 'a better and more respectable form'.[43] As a result, the Society established three classes of membership: the engineers, or full members; 'a class of Gentlemen ... under the denomination of honorary members'; and another category of honorary members consisting of 'various Artists ... connected with Civil Engineering'. Membership was increased to twenty-four, of whom twelve were elected in 1793. Sir Joseph Banks, President of the Royal Society, was elected as a Gentleman member, while Jesse Ramsden and John Troughton were recruited as Artists. Of the other ten listed as being elected between 1793 and 1800, only two had any engineering training - John Golborne and Sir Thomas Hyde Page, of whom the latter was a military engineer who had been trained at Woolwich. Another, Colonel William Mudge, could be regarded as a military engineer because he was the Director of the Ordnance Survey. Sir Samuel Bentham was elected as a Gentleman, and two - John Foulds, a millwright, and John Watte, a land surveyor - as Artists. One, Captain Joseph Huddart, a specialist in hydrographic surveys, could be regarded as the first marine engineer member, although properly speaking 'marine engineering' did not begin until the introduction of steam power to ships at the beginning of the nineteenth century. Another, William Chapman, went to sea as a boy but was enabled later by private means to decide on a career in civil engineering after correspondence with Smeaton.[44] Four were elected FRS: Page, Huddart, James Cockshutt, and Mudge. There were no obvious Scotsmen in this intake, but one, Chapman, was a Yorkshireman.

In summary, out of a total number of forty-two members listed in the records of the Society between 1771 and 1800, there was a modest but influential Scottish contingent of three compared with seven for Yorkshire, and the number of members who had also been fellows of the Royal Society totalled eleven, without including Banks. Very few came to the profession with any background of engineering, and those who did received it from family instruction, several important dynasties of engineers being represented amongst this early membership, or from working under people like Brindley and Smeaton. More frequently, members had been recruited from related professions such as land surveying, architecture, and military engineering, although a few millwrights achieved membership, the proportion declining sharply over the period. Very few of these early members had much familiarity with steam technology, apart from the outstanding exceptions of Boulton and Watt, and mechanical engineering generally was not strongly represented.

The Society of Civil Engineers never aspired to be a fully developed professional body for engineers. Smeaton modelled it partly on his experience of the Royal Society, so that it possessed a 'learned society' element from the outset, and it was also probably influenced by the contemporary vogue for gentlemen's dining clubs in London. Certainly, the fortnightly gathering for dinner during the parliamentary season became the most distinctive function of the Society, and it made no effort to provide the sort of services - protection of entry into the profession, definition of training standards, definition of professional standards, etc. - which have come to be expected of a professional association. Nevertheless, it is a mistake to dismiss the Society as being merely a dining club, as though it was of no importance.[45] There can be little doubt that the meetings were primarily social and convivial, but the fact that the leading British consulting engineers met regularly in congenial circumstances gave them the opportunity to discuss professional matters even though these were rarely recorded in the minutes. We know that they were at least occasionally discussed, because on one such occasion the debate generated some ill-feeling, and precipitated the reorganisation of 1793. It seems that a dispute occurred when the President, Joseph Nickalls, was rude to Smeaton, and although the former apologised quite handsomely the members took the opportunity to displace Nickalls by abolishing the office of President. Dr Wright has made the interesting suggestion that the cause of the dispute between Nickalls and Smeaton was a disagreement regarding the conflicting reports which they had given to the Bristol Society of Merchant Venturers regarding proposals to improve Bristol docks.[46] For the most part, such professional disagreements did not feature prominently in the business of the Society. A close reading of the first Minute Book of the Society, covering the years 1771-92, conveys a strong sense of the infrequency of the meetings and the irregularity of the attendance of the members. For example, at the fourth and last meeting recorded in 1774, the two members present, Pinchbeck and Jessop: 'Resolved - Nem Con by each of them all of them, and both of them for himself themselves and each and every one of them...' that circular letters be sent out urging members to attend more regularly.[47] But the very existence of the Society, and the value placed upon it by its membership of leading engineers, constituted it as an embryonic professional association and as a model for future developments.

An important feature of the process of rapid industrialisation in Britain in the eighteenth and nineteenth centuries is that it was by no means confined to the capital city, but was particularly strong in the provinces. In these circumstances, the search for the origin of professional consciousness amongst British engineers should not be limited to developments in London. The Society of Civil Engineers was a metropolitan group geared to the London parliamentary programme, and its members spent most of their working lives in the provinces, where they frequently took an active part in intellectual, social, and professional organisations. The history of the Lunar Society of Birmingham, for instance, is well known, and it is likely that James Watt would have received more stimulus to his amazing technological creativity from his colleagues in this provincial group than from the company of the Society of Civil Engineers.[48] The community of engineers assembled in the Soho enterprise by Boulton and Watt was, moreover, the most significant group of mechanical engineering expertise anywhere in the world at the end of the eighteenth century, with men such as William Murdoch, James Law, Isaac Perrins, John Southern,

James Lawson, Peter Ewart, Thomas Wilson, and William Brunton, all concerned with it at some time.[49] What was true of Birmingham, both in terms of 'philosophical' discussion and industrial practice requiring engineering skills, was true in varying degrees for other provincial centres like Manchester, Newcastle, Sheffield, Glasgow, Leeds, Liverpool, and Bristol. All were experiencing the animating effects of rapid industrialisation and expressing it through 'Literary and Philosophical Societies' or comparable organisations. In Manchester, great interest was expressed in the problems of designing large mills, improvements in textile machines, and in the application of steam power.[50] On Tyneside, the widespread introduction of Newcomen engines into the coal mines for pumping purposes increased engineering expertise and led to an interest in other aspects of mining technology, particularly in relation to making coal-mining safer as a result of better ventilation and lighting.[51] In Liverpool, the rapid increase in trade and consequent rise in the prosperity and activity of the port promoted continuing improvements in port facilities, and Henry Berry was employed to develop the system of enclosed docks begun earlier in the century by Thomas Steers.[52] In Bristol, a corresponding increase in port activity led to several engineers, including Nickalls and Smeaton, being commissioned to recommend dock improvements, but these were frustrated by indecision about the attractions of rival schemes until that of William Jessop was finally accepted in 1803.[53] Land drainage schemes continued to attract engineering attention, and all over the country the demand for better roads encouraged the formation of local Turnpike Trusts to improve existing road surfaces and to construct new roads by investing them with powers to raise funds from tolls charged on road users. These road building enterprises tended to make do with the minimum of professional advice, but gradually recognised the value of engineering expertise. The case of the Bath Turnpike Trust is not untypical in this respect. Formed as early as 1707, this was one of the first trusts for road improvement, but it was only after a reorganisation in 1756 that it gave serious attention to the construction of new roads, and then it tended to rely on the advice of architects such as C. Harcourt Masters of Bath, or local surveyors, rather than engineers. Only in 1818 did it appoint its first engineer, Benjamin Wingrove, and his background was that of a surveyor. Moreover, he was not a success, and it was only when the Trust appointed John Loudon MacAdam as its consultant engineer in 1822 that this particular enterprise acquired the services of a recognised engineer - although paradoxically MacAdam was always at pains not to describe himself as an engineer.[54]

In contrast with road building experience, the construction of canals required special skills from the outset, so that the embryonic engineering profession came to be intimately related to the boom in British canal building in the second half of the eighteenth century. The Duke of Bridgewater's first canal from his Worsley estates into Manchester, opened in 1761, required little special legislation (except for the Barton aqueduct over the River Irwell) because it ran mainly over his own property, and the engineering involved was comparatively elementary because there was no change of level. But the happy accident which led to the employment of James Brindley by the Duke brought a remarkable conjunction of talent with opportunity, for Brindley quickly demonstrated that he had a natural aptitude for the new types of engineering works necessary to lay out waterways, to make them water-tight, and to carry them over aqueducts and through hills by locks and tunnels. He demonstrated also a rare talent for exposition before parliamentary committees which combined incongruously with his rough-and-ready manner as a millwright. Virtually from the beginning, therefore, no substantial canal was undertaken without the proprietors seeking the professional advice of Brindley, Smeaton, or some other engineer, to lay out the line of their proposed waterway, to present the case to parliament for a special Act to allow it to be built, and to supervise the construction of the work once the Act had been passed. The relations between Brindley and his clients were usually obscure, with the clients tending to regard him as an employee rather than as a consultant. But Smeaton was more precise, and always defined a very clear consultancy arrangement with his clients, which left him free to undertake commissions as and when he wished to do so. Indeed, he soon took this relationship further and defined a hierarchy of engineering supervision, beginning with himself as consultant and passing through resident engineers, who were employed by the canal company, and such assistant engineers as were required for the project. Such a hierarchy was first established by Smeaton for the Forth and Clyde Canal, and was subsequently adopted by Rennie, Telford, and other engineers on all major works. In this way, canal construction served to give substance and form to the new profession.[55]

Canals were built in most parts of the country, from the remote Highlands of Scotland to the agricultural backwaters of Somerset and Devon, but they were most abundant in those Midland provincial areas which lacked access to natural waterways, so that they provided a great impetus to the industrialisation of Birmingham, the Black Country, the Potteries, and similar districts. One provincial area which had comparatively little need for canals, however, was nevertheless a notable repository of engineering talent: this was the County of Cornwall. Increasing industrial demand for copper and tin had caused a boom in the traditional mining of these minerals in Cornwall in the eighteenth century, and this had been an important factor in the development of the steam engine, first by Captain Thomas Savery, and then by Thomas Newcomen. In the event, however, the heavy fuel consumption of Newcomen's machine made it more applicable to coal mines, where fuel was abundant and cheap, than to the mines of Cornwall, where coal was a comparatively scarce commodity. The proprietors of Cornish mines continued to search for more efficient prime-movers, therefore, first welcoming the Watt engine and then becoming frustrated by the restrictions which the Watt patent set upon further improvements until it lapsed in 1800. The closing decades of the eighteenth century were thus a period of intense engineering activity in Cornwall. On the one hand, Boulton and Watt set up an extensive organisation in the county to construct, monitor, and service their engines, with brilliant agents such as William Murdoch carrying out his experiments with steam locomotion and coal-gas lighting while he worked there for the Birmingham firm.[56] On the other hand, Cornish native talent for engineering was expressed by the Hornblower family, rudely dismissed as 'Trumpeters' by Matthew Boulton, and by a number of young men like Richard Trevithick and Arthur Woolf who chafed at the restriction of the Boulton & Watt patent.[57] The Hornblowers had been amongst the first engineers to build Newcomen's engines, and had gone on to pioneer steam technology in America. They had taken root in Cornwall to become a thorn in the flesh of Boulton and Watt with their claims to produce machines which were as efficient as those from

Birmingham. This Cornish engineering community remained unorganised in any formal
sense, but it flourished in the first half of the nineteenth century with the development
of high pressure steam, compounding, and the superb Cornish engines which were to set
a pattern for the rest of the world.[58]

All of this provincial engineering activity influenced metropolitan developments in
one way or the other. Not only did it cause the leading engineers to make their regular
visits to London in order to present the cases of their clients to parliament, although, as
we have seen, this was exceptionally important in the creation of the Society of Civil Engineers. It also helped to establish other linkages. For example, Thomas Yeoman, the
first President of the Society, had been an active member of the Northampton Philosophical Society before his move to London in 1757, and he went on to become a prominent
member of the Society of Arts, formed in 1754, to promote enterprise and innovation by
the award of annual prizes. Yeoman was Chairman of the Society of Arts Committee on
Mechanics from 1763 to 1778.[59] This increasing integration of provincial and metropolitan life provided fertile ground for the growth of professional consciousness amongst engineers.

Three general observations are appropriate on the emergence of the British engineering profession in the eighteenth century. First, it is clear that the total number of engineers concerned in this development remained small. The Society of Civil Engineers
only had forty-two members from its foundation in 1771 up to the end of the century. It
would seem adequate to allow for something like 150 resident engineers, assistants, and
pupils, associated with these members on their various projects, who could be regarded
as likely to move up to full professional status. A less certain addition, but probably not
more than a few dozen, would allow for the senior freelance engineers who thrived in the
provinces, especially as mechanical engineers, but also in land drainage and the service
of large estates. This would give an estimated total of potential professional personnel in
the second half of the eighteenth century of only about 260. Of course, there was a large
number of millwrights, stonemasons, instrument makers, and other craftsmen, capable
of being promoted into this select group given the right conjunctions of favourable circumstances. But they need not be included amongst the practising professional engineers of the period.

The second general observation is that this group of proto-professional engineers
were a very motley crew. They came from a wide variety of backgrounds, with different
sorts of training - even with no particular training at all. Most of them began their careers as some type of craftsmen: millwright, stonemason, instrument maker, and such like.
A tiny handful of them had received some kind of professional experience through training as lawyers, like Smeaton, or as architects, like Mylne, but these 'gentlemen engineers'
were of outstanding importance because they mediated their professional experience to
the group of engineers which they established. Such diversity remained a dominant characteristic of the British engineering profession until well into the nineteenth century, and
served to provide a barrier against the standardisation of entrance requirements .

Despite both the paucity of numbers and the diversity of background, however, the
third observation is that the engineering profession in Britain certainly emerged in the
second half of the eighteenth century. Several important engineers were around in the

1750s, but the act of creation was virtually the single-handed work of John Smeaton, who stepped into the still inchoate world of civil engineering and assumed professional status for it and for himself. In doing so, he created the profession in his own image, as one suitable for gentlemen, modelled on the Royal Society in its institutional forms, with an established hierarchy of consultancy-client relationships, and with a clear recognition of its professional character. These qualities were incorporated by Smeaton in the Society of Civil Engineers which, although undeniably social in its primary function, became an active professional association by virtue of its membership and the regularity of its meetings. Needless to say, it did not perform all the functions expected of a modern professional body, and, in particular, it made no provision for the training of entrants to the profession. This remained a matter of personal initiative: a man demonstrated that he was an engineer by his works, and when he had demonstrated it successfully he was admitted to the professional body - a self-validating procedure which was to play a large part in the development of professional engineering in Britain.

By 1800 one can speak with confidence of a small but effective engineering profession in Britain. It had clearly drawn apart from the more rudimentary crafts out of which it had largely sprung, and from neighbouring and potentially rival professional groups such as the architects and surveyors. The separation from the surveyors was the last to occur, but surveying was a comparatively simple operation, and was frequently performed by a person such as William Smith, the canal builder and mining engineer, who described himself both as a surveyor and an engineer long before he became recognised as the 'father of English geology'.[60] The role of surveyor, indeed, tended to be subordinated to that of engineer, especially after the development of the Ordnance Survey deprived the surveyors of their important cartographic function, and they did not achieve the professional status of possessing an Institution of their own until 1868.[61] But so far as the engineers were concerned, the opening of the new century presented them with almost unlimited prospects for progress, and they had achieved the self-consciousness necessary to enable them to seize these opportunities. The editor of Smeaton's Reports, published in three volumes by the Society in 1812, spoke perceptively of 'a new era in all the arts and sciences' beginning around 1760, when Smeaton's influence began to impress itself on the profession.[62] Similarly, a generation later, the new Institution of Civil Engineers honoured Smeaton by incorporating the Eddystone Rock Lighthouse as the dominant symbol in its coat-of-arms. It was a fitting recognition for an exceptional personal achievement.

Notes

1. The traditional usage was first established by Arnold Toynbee: *Lectures on the Industrial Revolution in England*, London, 1884, and subsequently adopted by scholars such as Paul Mantoux, *The Industrial Revolution in the Eighteenth Century* translated from the French, London, 1928 (Jonathan Cape revised edition, 1961); and T. S. Ashton, *The Industrial Revolution 1760-1830*, Oxford, 1948. Amongst more recent discussions which have questioned the limitations of period, see R. M. Hartwell (Ed.), *The Causes of the Industrial Revolution in England*, Methuen, London, 1967; and David S. Landes, *The Unbound*

Prometheus - technological Change and Industrial Development in Western Europe from 1750 to the Present, Cambridge, 1969

2. J. Y. Nef: 'The Progress of Technology and the Growth of Large Scale Industry in Great Britain, 1540-1640', in *Econ.Hist.Review*, 1st series, 5, 1934, has argued for an earlier date to accommodate the sixteenth century developments in industry and commerce; and E.M. Carus-Wilson: 'An Industrial Revolution of the Thirteenth Century', *Econ.Hist.Review*, 1st series, 11, 1941, found medieval precedents. Both these articles were reprinted in E. M. Carus-Wilson (Ed.), *Essays in Economic History*, Volume 1, Edward Arnold, London, 1954.

3. W. W. Rostow, *The Process of Economic Growth*, Oxford, 1960, has been particularly useful in promoting the generalisation of Industrial Revolution; Donald N. McCloskey (Ed.), *Essays on a Mature Economy - Britain after 1840*, Methuen, London, 1971, shows a further extension of the concept by modern econometric historians.

4. L. T. C. Rolt and J. S. Allen, *The Steam Engine of Thomas Newcomen*, Moorland, Hartington, 1977; Carroll W. Pursell Jr., *Early Stationary Steam Engines in America - A study in the migration of a technology*, Smithsonian, Washington D.C., 1969.

5. Landes, *op. cit.* p.96.

6. George W. Daniels, *The Early English Cotton Industry*, Manchester, 1920, gives a convenient account of these innovations.

7. For a recent study of the development of a major port, see Nancy Ritchie-Noakes, *Liverpool's Historic Waterfront - The World's First Mercantile Dock System*, HMSO, London, 1984.

8. J. R. Ward, *The Finance of Canal Building in Eighteenth Century England*, Oxford, 1974. See also R. A. Buchanan, 'The British Canal Engineers, The men and their resources' in Per Sörbom (Ed.), *Transport Technology and Social Change*, Tekniska Museet Symposia, Stockholm, 1980, pp. 67-89.

9. There is a large body of scholarly literature on the British Agricultural Revolution. For its close relationship with the Industrial Revolution see especially E. L. Jones, *Agriculture and the Industrial Revolution*, Basil Blackwell, Oxford, 1974.

10. Landes, *op. cit.*, Chapters 1 and 2.

11. Georgius Agricola, *De Re Metallica*, Basle, 1556, translated by H. C. Hoover and L. H. Hoover and available in the Dover edition, New York, 1950. See also M. B. Donald, *Elizabethan Monopolies - A History of the Company of Mineral and Battery Works*, Edinburgh and London, 1961.

12. S. Smiles, *Lives of the Engineers*, volume 1, has six chapters on Hugh Myddelton. He describes James I conferring a baronetcy on Myddelton in 1622 (p. 143), after a knighthood in 1613 (p. 125). But in a later edition he dismissed the 'usual statement' that Myddelton was knighted on the opening of the New River in 1613 as inaccurate.

13. Smiles, Ibid. chapter III, p. 43: Vermuyden 'received the honour of a knighthood at the hands of Charles I' on 6th January 1629.

14. Smiles, Ibid. See also: H. C. Darby, *The Drainage of the Fens*, Cambridge, 1940, 2nd edition, 1956, reprinted 1968; Richard L. Hills, *Machines, Mills and Uncountable Costly Necessities. A short history of the drainage of the Fens*, Goose and Son, Norwich, 1967; and L. E. Harris, *Vermuyden and the Fens - A study of Sir Cornelius Vermuyden and the Great Level*, Cleaver-Hume Press, London, 1953.

15. L. E. Harris, Ibid, p. 98. *The Concise DNB* gives '1683?' as Vermuyden's date of death.

16. See L. T. C. Rolt and J. S. Allen, *op. cit.* (note 4 above), pp. 31 and 35.

17. Ibid. p. 35, and personal communication with Mr. J. S. Allen.

18. Ibid. chapters 4, 5 and 6.

19. The relationship between Protestant nonconformist sects and business interests is not well documented for the Baptists. But for a close parallel group, see Arthur Raistrick, *Quakers in Science and Industry*, David and Charles, Newton Abbot, 1968 (first published in 1950).

20. There is a useful review of the part played by these men in Rolt and Allen, *op. cit.* On Triewald and the process of transmitting steam technology overseas, the literature has recently been greatly enriched by Svante Lindqvist, *Technology on Trial - The Introduction of Steam Power Technology into Sweden, 1715-1736*, Uppsala, 1984.

21. The most succinct account of Brindley's career is still S. Smiles, *Lives of the Engineers*, vol. 1, pp. 307-346. But see also Cyril T.G. Boucher, *James Brindley Engineer 1716-1772*, Goose, Norwich, 1968.

22. William Fairbairn, *Treatise on mills and millwork*, London, 1878, Preface to the first edition, 1861.

23. Information about the Hornblower family is somewhat scattered. There is, however, a full genealogy at the end of William Nelson, 'Josiah Hornblower and the first steam-engine in America', in *Proceedings of the New Jersey Historical Society*, 2nd series 7, 1883, 177-247. The other engineers mentioned are all discussed below.

24. Charles Hadfield and A.W. Skempton, *William Jessop, Engineer*, David and Charles, Newton Abbot, 1979, p. 12: 'In 1759 Smeaton took William as an apprentice, to serve a 'clerkship', before becoming his draughtsman and assistant.' After Jessop's father died in 1761, his premiums were paid to Smeaton by Robert Weston, a leading promoter of the Eddystone Lighthouse re-building.

25. Joseph Mitchell, *Reminiscences of my life in the Highlands,* published privately 1883, reissued by David and Charles, Newton Abbot, 1971, 2 vols. Mitchell describes Telford's London establishment in vol. 1, pp. 87-8.

26. See L. T. C. Rolt, *Tools for the Job - A Short History of Machine Tools*, Batsford, London, 1965.

27. Whitworth Porter, *History of the Corps of Royal Engineers*, London, 1889, 2 vols.

28. Ibid. vol. 1, p. 167. The establishment of the Royal Military Academy at Woolwich in 1741 was an important development here, initiating 'a process of formal training for the technical services', see D.W. Marshall, *The British Military Engineers 1741-1783, A study of organisation, social origin and cartography*, unpublished PhD dissertation, University of Michigan, 1976, p. 87.

29. Porter, *op. cit.* vol. 2, p. 169.

30. For accounts of Wade and his roads, see J.B. Salmond, *Wade in Scotland*, Moray Press, Edinburgh and London, 1934; and William Taylor, *The Military Roads in Scotland*, David and Charles, Newton Abbot, 1976.

31. For Sorocold, see F. Williamson, 'George Sorocold of Derby - A pioneer of water supply', in *Journal Derbys. Archaeological and Nat. Hist. Soc.* 1936, 43-93: 'The first man outside London to install "house-to-house" water supply, and in his time had a great reputation as an engineer' (p. 43). On Liverpool docks in the eighteenth century, see Henry Peet, 'Thomas Steers - The Engineer of Liverpool's First Dock - A Memoir' in *Trans. Hist. Soc. Lancs. and Cheshire*, 82, 1930, 163-242; and Stanley A. Harris, 'Henry Berry (1720-1812): Liverpool's Second Dock Engineer', in *Trans. Hist. Soc. Lancs. and Cheshire*, 89, 1937, 91-116.

32. Professor Skempton gives 1768 as the year when Smeaton 'first described himself as a "civil engineer" on the title page of one of his printed reports' - see A.W. Skempton (Ed.), *John Smeaton FRS*, Thomas Telford, London, 1981, p. 4.

33. A.W. Skempton and Esther Clark Wright, 'Early members of the Smeatonian Society of Civil Engineers' in *Trans. Newcomen Soc.*, 44, 1971-2, 22-47, p. 25. See also Eric Robinson, 'The Profession of Civil Engineer in the Eighteenth Century - A Portrait of Thomas Yeoman FRS, 1704-1781', in A. E. Musson and Eric Robinson, *Science and Technology in the Industrial Revolution*, Manchester, 1969, chap. 9, pp. 372-392; and A. W. Skempton (Ed.), *John Smeaton FRS*, 1981, p. 233.

34. Smeaton wrote his own detailed account of the lighthouse, see J. Smeaton, *A Narrative of the Building and a Description of the Construction of the Edystone (sic) Lighthouse*, London, 1791. See also Rowland Mainstone, 'The Eddystone lighthouse', in A.W. Skempton (Ed.), *John Smeaton FRS*, 1981, chap. 4, pp. 83-102.

35. A principle established in the Wells Harbour Lawsuit, 21 November 1782 (Michaelmas Term, 23 Geo III) in 3 Douglas's Reports, King's bench (1782), pp. 157-161, Folkes/Bart v. Chadd and others. See Denis Smith, chap. 10 'Professional Practice' in A. W. Skempton (Ed.), *op. cit.*, pp. 217-227, especially pp. 226-7.

36. 'At first Smeaton thought the Watt engine too complicated, and that its construction made too great a demand for the engineering skills of the day' - H. W. Dickinson and Rhys Jenkins, *James Watt and the Steam Engine*, Oxford, 1927, p. 300. But he was soon convinced of the superiority of performance of the Watt engine design and appears to have relinquished work on new engines to Boulton and Watt, even though 'It seems ... that he continued to be concerned in the construction of atmospheric engines for collieries after this date' (Ibid.)

37. The poet Robert Southey referred admiringly to Brindley's bluff but effective presentation of his case to official bodies in *Espriella's Letters from England*, 1807. But the fullest account is that of Smiles, *Lives, op. cit.* vol. 1.

38. L. St. L. Pendred, *British Engineering Societies*, Longman, London, 1947, p. 11, refers to them as friends; F. M. L. Thompson, *Chartered Surveyors: the growth of a profession*, Routledge, London, 1968, p. 60 describes them as 'great rivals'. On occasions they certainly gave conflicting advice, as when reporting to the Committee of the Forth and Clyde Canal in 1768, described by Anthony Burton, *The Canal Builders*, Eyre Methuen, London, 1972, p. 95.

39. Denis Smith, 'The Professional Correspondence of John Smeaton: an eighteenth-century consulting engineering practice', in *Trans. Newcomen Soc.*, 47, 1974-6, 179-89, especially 182.

40. Skempton and Wright, *op. cit.* See also R. A. Buchanan, 'Steam and the Engineering Community in the Eighteenth Century' in *Trans. Newcomen Soc.*, 50, 1978-9, 193-202, on which I have drawn extensively in this section.

41. This analysis is based only on the members named by Skempton and Wright, and does not include other members who were largely 'gentlemen' members.

42. Skempton and Wright, *op. cit.*, p. 30.

43. Ibid. p. 37.

44. Chapman was elected to the Society of Civil Engineers in 1795. As a young man he had worked in the Newcastle district in collieries leased by his family. See Skempton and Wright, *op. cit.*, p. 39, and also A. W. Skempton, 'William Chapman (1749-1832), Civil Engineer', in *Trans. Newcomen Soc.*, 46, 1973-4, 45-82.

45. For a recent example see Göran Ahlström, *Engineers and Industrial Growth*, Croom Helm, London, 1982, p. 86.

46. S. B. Donkin, 'The Society of Civil Engineers (Smeatonians)' in *Trans. Newcomen Soc.*, 17, 1936-7, 51-71: Dr. Wright made a contribution to the discussion on this paper containing the suggestion - 'About 1786 Smeaton recommended Nickalls and Jessop to the Committee of the Society of Merchant Venturers of Bristol ... Finally, however, Smeaton was called to Bristol, and he condemned both plans. Hence it is not surprising if at about that time there had been some friction between the founder and the President of the Smeatonian Society'.

47. Minute Book 1771-92 of the Society of Civil Engineers, 27 May 1774. Having passed this verbose resolution while presumably enjoying their convivial occasion, the two members adjourned until the next Parliamentary session, and the next recorded meeting is on 8 March 1776. Apparently there were no meetings at all in 1775. Christopher Pinchbeck is a shadowy figure who is not mentioned in Skempton and Wright, *op. cit.* But the minutes show that he was elected into membership on 29 April 1774 and that he was President of the Society from 15 May 1781 to his death on 25 March 1783.

48. See Robert E. Schofield, *The Lunar Society of Birmingham - A Social History of Provincial Science and Industry in Eighteenth-century England*, Oxford, 1963.

49. Dickinson and Jenkins, *op. cit.*, especially chap. XXI, 'The Staff of Boulton and Watt'.

50. Musson and Robinson, *op. cit.*, especially chap. XII, 'The Early Growth of Steam Power'.

51. See A. Raistrick, 'The Steam Engine on Tyneside, 1715-1778', in *Trans. Newcomen Soc.*, 17, 1936-7, 131-164.

52. See above, note 31.

53. See R. A. Buchanan, 'The Construction of the Floating Harbour in Bristol: 1804-1809' in *Trans. Bristol and Gloucestershire Archaeological Society*, 88, 1969, 184-204; also Alan F. Williams, 'Bristol Port Plans and Improvement Schemes of the Eighteenth Century' in *Trans. BGAS*, 81, 1962.

54. See A. Gibb, *The Story of Telford*, 1935, pp. 178-9, where the account of rivalry between MacAdam and Telford is tinged with professional jealousy. But MacAdam preferred to describe himself as an 'Inspector of Roads' or 'Surveyor of Roads': see the recent biography by W.J. Reader, *MacAdam - The MacAdam Family and the Turnpike Roads 1798-1861*, Heinemann, London, 1980. See also B.J. Buchanan, 'The Evolution of the English Turnpike Trusts: Lessons from a Case Study', in *Economic History Review*, 2nd series, 39, 1986, 223-43.

55. For Smeaton, see Denis Smith, 'Professional Practice', chap. 10 in A. W. Skempton (Ed.), *John Smeaton FRS*, 1981, pp. 217-227, especially pp. 224-225. Burton, *op. cit.*, pp. 127-8, quotes Rennie's reports on the Kennet and Avon canal to show how he regarded this structural differentiation. See also Kenneth R. Clew, The Kennet and Avon Canal, David and Charles, Newton Abbot, 1968, p. 51; and C.T.G. Boucher, *John Rennie, 1761-1821: The Life and Work of a Great Engineer*, Manchester, 1963, pp. 63-5

56. Dickinson and Jenkins, *op. cit.*, pp. 290-7, give a useful summary of Murdoch's career.

57. Dickinson and Jenkins, *op. cit.*, chap. XXII, 'Rivals and Pirates', 298-327. see also J. Hambley Rowe, 'The early history of Hayle Foundry 1770-1833' in *Royal Cornwall Polytechnic Soc.* n. s. 8 (3) 1936, 40-9; and A. K. Bruce, 'Boulton and Watt v. Bull' in *The Engineer*, 29th September 1944, 240-1

58. Dickinson and Jenkins, *op. cit.*, 303-8. see also Thomas Lean, *Historical Statement of the Improvements made in the duty performed by the Steam Engines in Cornwall*, 1839, reprinted by Bradford Barton, Truro, 1969; G. N. von Tunzelmann, 'Technological Diffusion during the Industrial Revolution' in R. M. Hartwell (Ed.), *The Industrial Revolution*, Oxford, 1970; W. T. Hooper, 'Perran Foundry and its story' in *Royal Cornwall Polytechnic Soc.*, n.s. 9, 3, 1939, 62-89; and Edgar Loam, 'Michael Loam, Engineer', in *Royal Cornwall Polytechnic Soc.*, 7, 1932, 142-58

59. Skempton and Wright *op. cit.*, p. 24. also see Musson and Robinson *op. cit.*, chap. XI. For the Society of Arts, see Sir Henry Trueman Wood, *A History of the Royal Society of Arts*, John Murray, London, 1913. The founder of the Society was William Shipley, a friend of Yeoman in Northampton .

60. William Smith has been the subject of several short memoirs, but there is as yet no substantial work on him. See Mrs J. M. Eyles' article on Smith in the *Dictionary of Scientific Biography*, New York, 1975, XII.

61. See F. M. L. Thompson, *op. cit.*, published to celebrate the first centenary of the Chartered Surveyors in 1968.

62. John Smeaton, *Reports ... made on various occasions in the course of his employment as a Civil Engineer*, 3 vols, London, 1812, Preface to 1, p. iii. The Preface was probably written by Mylne, who did much of the editing of this work for the Society of Civil Engineers. Strangely, however, I have been unable to find any reference to the adoption of the title 'Smeatonian Society of Civil Engineers' in the Minute Books of the Society: S. B. Donkin, *op. cit.*, p. 68, also confessed his inability to put a date on this event.

Chapter Three

The Growth to Maturity

The foundation of the Institution of Civil Engineers in 1818 was a landmark in the history of the British engineering profession, representing the establishment of a secure organisational base for a self-conscious group of engineers which was able to grow quickly into a mature association and to become the exemplar of professional organisation for engineers thereafter. It was not the first organisation of professional engineers, for as we have seen the Society of Civil Engineers had fulfilled an important pioneering function even before its re-creation in 1793, and had attracted most of the senior engineers of the time into its membership. And it was certainly not the last such organisation, for there was a remarkable proliferation of engineering associations from the middle of the nineteenth century, with many new national institutions being formed as well as new regional societies of engineers and similar organisations overseas. One of the most striking things about all these later foundations, however, was the regularity with which they turned to the Civils as their model, and frequently for direct advice and even for the loan of premises. Wherever engineers have banded together in any part of the world, they have drawn inspiration and guidance from the experience of the Civils. In the most constructive sense, therefore, the British Institution of Civil Engineers has been an archetype or prototype for subsequent organisation, so that it is particularly important to understand how it came to be regarded as such a significant creation. In this chapter we will consider first the background against which it took shape, in the wake of the great vogue for canal-building in Britain. Then we will examine the seminal early years of the Civils and seek to explain why the Institution came to exert such a powerful influence on the engineering profession.

The biggest problem of engineering history in this period is why there was so little continuity between the Society of Civil Engineers and the Institution of Civil Engineers. The claim for the Society to be recognised as the first formal association of professional engineers has already been demonstrated. It is beyond reasonable doubt that John Smeaton's vision of the role of civil engineering established him as the 'Father' of his profession, and that his Society of Civil Engineers set a pattern of organisation which was a valuable precedent for subsequent bodies. Yet by the beginning of the nineteenth century, the Society had already become remote from the interests of most of the rising young engineers seeking to establish their reputations in canal and harbour works, machine-making and land drainage. It seems that membership had become exclusive, consisting of an elite of senior canal engineers who were not eager to recruit younger engineers into what had become, in effect, a select dining club. The result was that it offered little en-

couragement to the next generation of engineers, who were left largely to their own resources in the expression of professional consciousness. Fortunately for them, they found
an ally and a leader amongst the highest ranks of British engineers in the person of Thomas Telford, who had his own reasons for disliking the elite of the Society of Civil Engineers. To understand this relationship it is necessary to review the development of
engineering in Britain from the 'Canal Mania' of the 1790s.

Perhaps the most striking feature of canal-building in Britain is the shortness of its
duration.[1] Excluding such long-standing artificial water-ways as the Exeter Canal, and
the river improvements which had been made in the first half of the eighteenth century,
the first genuine canals were begun in the 1750s with the Sankey Navigation (1755) and
the original Bridgewater Canal from Worsley to Manchester (1759). To some extent, the
Newry Canal in Ireland (1730) had provided a useful introduction to the new technique,
but it did not immediately establish a pattern of canal-building. It was rather the engineering of James Brindley under the inspired entrepreneurship of the Duke of Bridgewater and his agent John Gilbert which provided the impetus behind British canal
construction. When the Duke died in 1803, leaving estates made rich by a lifetime devoted
to canal work, the main period of canal construction was already over. True, there were
many projects, some of them delayed by the French wars, in the course of completion,
and a number of significant improvements to the canal system remained to be done. But
with the passing of the 'Canal Mania' of the mid-1790s, the number and scale of canal
operations fell off steadily until work ceased altogether in the 1840s. The only important
canal to have been built in Britain since then has been the Manchester Ship Canal (1885-
1894), and that deserves to be considered as an extension of port activity rather than as
a traditional inland waterway.[2]

Within this remarkably short time span, canal engineering fell conveniently into three
main periods. First, the years of the early canals, from the 1750s to the 1780s, were dominated by the Duke of Bridgewater and his engineer James Brindley, and to this period
belong the beginnings of twenty-two major waterways. The second period covers the
boom years otherwise known as the 'Canal Mania', which lasted for less than a decade
from the late 1780s to the mid-1790s. In the flurry of canal speculation which characterised these years, thirty-five ventures were commenced in England, all seven of the
main Welsh canals, the Crinan Canal in Scotland and the Royal Canal in Ireland. The
third period followed the collapse of the 'Mania', running from 1796 to the cessation of
British canal construction in the 1840s. Twenty-two canals were built in England in these
years, and four - including the Caledonian Canal - in Scotland, but there were no additions to the Welsh and Irish systems except for the Ulster Canal.

As an engineering achievement, the British Canal Age was typical of the Industrial
Revolution, to which it made an important contribution, in its parochialism. British canal
building was a distinctly homespun affair. There had been canals in Italy, France, Holland and elsewhere long before Brindley puddled his first trench in Lancashire, but apart
from the inspiration which such works may have given the young Duke of Bridgewater
during his Grand Tour, there is no evidence of this continental expertise having much influence on British practice. This is somewhat surprising, because there had been plenty
of continental involvement in British mining and land drainage engineering during the

sixteenth and seventeenth centuries, so that the notion of calling in foreign advisers when novel and difficult feats of engineering were being undertaken was not unfamiliar to British entrepreneurs. There is some evidence that Rennie, Telford and others took note of French theoretical work: Telford is reputed to have learnt French in order to study Belidor. But when it came to constructing canals in Britain, only British engineers were involved.[3] The proprietors of the Sankey Navigation called in Henry Berry, the Liverpool Docks Engineer, to advise them, and it was Berry's predecessor at Liverpool, Thomas Steers, who had designed the Newry Canal. Liverpool was the first British port to employ an engineer on a regular basis, and it thus provided a natural choice for any other venture in the North West requiring engineering advice.[4] Thereafter, British canal engineering expertise was largely self-generated, and certainly no foreign engineers ever became involved in it to any significant extent.

Conversely, there is little evidence of British canal engineering expertise being carried elsewhere, unlike the subsequent boom in railway engineering which sent British engineers out to all corners of the world in the process of railway construction. The only significant exceptions were the involvement of Thomas Telford in the Götha Canal and the participation of some British engineers in North American canals, even though, in the latter case, the engineers concerned had not had much experience of British canal building. William Weston, who advised various groups of canal proprietors in Pennsylvania, Massachusetts, and New York, was virtually unknown in British practice, and Lt. Colonel By, who undertook the superb works of the Rideau Canal in Canada, was a military engineer for whom strategic considerations were dominant and who had no British experience of canal building.[5] The Canal Age in Britain was thus not only short; it was also very insular. However great the eventual ramifications of British canal building - and they were certainly immense - the canals were, in the first instance, designed by Britons for local British purposes, and they drew only on domestic experience and expertise.

It may reasonably be argued that the self-generated quality of British canal engineering was made possible by the essentially elementary nature of much canal construction. Fundamentally, the skill consists of the ability to build a level, watertight ditch, and this was a skill which had been acquired and practised over many generations of surveying and levelling for land drainage schemes and other works requiring experience of water control. That is to say that the basic skill of the canal engineer was available in the large body of people who, as surveyors or land agents or in other capacities, had learnt how to take levels, to make excavations and embankments and to provide a watertight medium by the use of puddled clay. Even the more sophisticated skill of providing a means of changing levels through pound locks had been acquired in Britain by the river navigation engineers in the decades immediately before the Canal Age, so that any trace of continental influence in this respect had already been mediated through a generation of British practice.[6] These basic skills of the canal engineer were thus part of the inheritance of the men who first applied themselves to the problems of canal building, and there was nothing more that could not be better learnt than by putting these skills into practice.

Once the operation of canal building had started, new solutions began to appear to the new breed of canal engineers, and especially to James Brindley, whose most notable innovations in canal practice were the masonry aqueduct and the canal tunnel. Thus equipped, the canal engineers of the great boom years were able to tackle almost every contingency, and in the process they developed some important refinements of technique. The most important of these was probably the introduction of cast iron as a major structural material, particularly in aqueducts such as the work of Jessop and Telford at Pont Cysylte and Chirk. Other innovations included the use of inclined planes and lifts as alternatives to pound locks in changing levels (these were particularly important in some of the later canals, where large changes of level were tackled with inadequate supplies of water for the traditional pound locks), and the development of the arched masonry dam (best exemplified by Lt. Colonel By's dam at Jones' Falls on the Rideau Canal in Canada). Little use, in fact, was made of these later innovations on British canals: the only viable canal lift was installed at Anderton in Cheshire in 1875, although the technique was speedily improved elsewhere, and the British inclined planes were usually small and had short lives. The techniques of the British canal builder remained fairly simple, and as such did not require sustenance from foreign expertise.[7]

British canal engineering was thus a short-term operation using fairly simple techniques which were available from home-spun talent, but while it lasted it called for an unprecedented degree of engineering organisation and it was for this reason that the canals did more than any other innovation to create the British engineering profession. Before 1760, there had been many surveyors able to measure estates and to plot the alignment of roads; there had been millwrights who were able to build steam engines and to undertake a wide variety of constructional tasks; there had been architects and masons ready to design bridges, and other craftsmen prepared to build harbours, waterworks and lighthouses, or to make clocks, instruments, and other complicated machines. But it was the coming of the canals which provided, for the first time, the financial resources for organising substantial employment for large bodies of people engaged in constructional activities, and this incentive converted the collection of individuals with the diverse talents and skills which have been mentioned into the beginnings of a genuinely professional group - the civil engineers.[8]

The transformation was never more plain than it was in the career of the itinerant millwright who became the first of the great canal engineers - James Brindley. Attention has already been drawn to Brindley's humble origins. Despite these, he quickly became known to an ever-widening circle of entrepreneurs and industrialists in the North Midlands who consulted him about drainage problems, waterworks, mills, machinery and steam engines, in dealing with all of which he showed striking ingenuity. Then in the late 1750s, Earl Gower consulted him about the possibility of constructing an artificial waterway linking the Trent with the Mersey, and although the project was not pursued immediately it served to bring Brindley to the attention of the young Duke of Bridgewater who, in 1759, was embarking on his career as a canal promoter by seeking to link his coal mines at Worsley with Manchester. At the age of forty-three, therefore, Brindley became involved in the work which was to occupy the remaining thirteen years of his life, and to win him immortality as the first engineer of the British canal system.[9]

For all his unquestioned eminence as an expert on canals from 1759 to his death in 1772, it remains difficult to see Brindley as a professional engineer in any conventional sense. He rarely seems to have charged professional fees. He received only a pittance for his work on the Bridgewater Navigations, according to the accounts which he kept in his own idiosyncratic way in his pocket book. Smiles attributed this to the precarious financial situation of the Duke of Bridgewater's affairs at this time, and suggests also that the lack of any significant financial dependence accounts for Brindley's striking independence in his relationship with the Duke. This may well be, but Brindley does not appear to have charged any of his 'clients' what would later have been regarded as a consultancy fee. Instead, he seems to have accepted commissions to build canals, as any millwright might take on jobs, and only under pressure of a multitude of such commitments to have admitted the need to delegate work to subordinates. Such delegation did not please canal proprietors, like those of the Coventry Canal who actually dismissed him because they understood that they were hiring a man to do a job, and objected to him spending time on other projects. On the other hand, the arrangement provided a superb training school for those subordinates who were left to get on with the tasks under Brindley's occasional but illuminating supervision, so that a generation of British canal engineers was raised in these years which was trained and inspired by James Brindley. While Smeaton was establishing a formal distinction between a 'consultant' engineer and the 'resident' engineer carrying out his plan, the difference in function appeared clearly in practice with people like Thomas Dadford, Robert Whitworth, Hugh Henshall and Josiah Clowes serving a sort of informal tutelage under Brindley's restless guidance.

Another way in which Brindley never quite fulfilled the stereotype of a professional man was in his public image. He was an individualist to the point of eccentricity, often preferring the company of his favourite mare to that of his human acquaintances. He disliked expressing his opinions in written reports or letters, preferring a face-to-face encounter. He did not care for London, and was shattered by his one and only visit to the theatre. Although he was clearly a most effective witness before parliamentary committees - because so many anecdotes of his skill with 'visual aids' survive, and his cases were usually successful - he did not adopt the conventional forms of professional behaviour on these occasions, and one is bound to wonder whether or not the better educated and more conventional Smeaton was influenced by this in not giving Brindley an invitation to join the Society of Civil Engineers which he established in 1771, the year before Brindley's death. It remains possible, however, that Brindley received such an invitation but rejected it; we know so little about the early history of the Society. Certainly, several of the engineers trained by Brindley became early members, including Whitworth, Dadford, and Henshall. But the suspicion remains that there was a taint of uncouthness about Brindley which rendered him unacceptable in the company of polite gentility. To this extent, he never quite achieved 'professional' status.[10]

The predominance of James Brindley in the first period of British canal construction, from 1755 to 1785, is amply demonstrated by the fact that, out of nineteen canals in England in this period, he was responsible for eight of them, at least until his death in 1772. Even in this period, however, other engineers were active in designing and building canals. The work of Liverpool Docks engineers, Steers and Berry, on the Newry and Sankey Canals respectively, has already been mentioned. The Gilbert brothers, in addition to being agents to the Duke of Bridgewater and Earl Gower, also ventured into canal engineering in their own right at Donnington Wood. John Grundy and Thomas Yeoman, land drainage engineers and early members of the Society of Civil Engineers, were active in the Eastern Counties and Gloucestershire respectively, and John Longbotham pioneered the Leeds and Liverpool Canal although it was not completed until long after his period as engineer. John Smeaton was consulted about several of the early canals, and became Chief Engineer of the only Scottish project in this period, the Forth and Clyde Canal. Smiles speculated that it could have been Smeaton who gave the unfavourable report on Brindley's scheme for the Barton aqueduct, saying: 'I have often heard of castles in the air, but never before saw where any of them were to be erected'.[11] Whether or not it was Smeaton who said this, there are clear indications that the professional opinions of the two men did not always coincide[12] and it is clear that Smeaton brought to canal construction the same sort of capacity for systematic organisation which had been so triumphantly vindicated by the completion of the Eddystone lighthouse in 1759. It was this quality which justified Rolt's judgment that: 'Smeaton found engineering a trade and made it a profession'.[13] The process of transformation which had been begun by Brindley was thus completed by Smeaton. This first generation of canal engineers worked out a system of professional organisation which made it possible for their successors to take the great expansion of the Mania years in their stride.

The Canal Mania, the second period of British canal engineering, lasted only from the late-1780s to the mid-1790s. It was dominated by a triumvirate of outstandingly able civil engineers of whom the senior member was William Jessop, with John Rennie and Thomas Telford as the two younger men who rose to prominence in these years. Until very recently, William Jessop (1745-1814) has been regarded as a shadowy figure, receiving less credit than he has deserved for the engineering achievements of the Mania period and the subsequent years of dock and tramway construction, in all of which he played an important part. Any injustice on this score has now been handsomely rectified in the new biography of Jessop by Hadfield and Skempton.[14] As a person, Jessop remains a rather elusive character, because of the loss of virtually all his personal papers, but his two biographers have done a commendable job in bringing together an account of Jessop's activities which demonstrates conclusively his predominance as a canal engineer in the Mania years. 'On his shoulders fell the weight of the canal mania', they observe, pointing out elsewhere that: 'Because he was the premier engineer of the day, most promoting committees tried first to get Jessop to agree their plans and help their Bill through Parliament'.[15] Jessop had been born in Devon and had the good fortune to be apprenticed to John Smeaton, after his father Josias had been employed by Smeaton to superintend work on the Eddystone lighthouse. The young Jessop moved to Smeaton's home at Austhorpe, Leeds, where he became the older man's assistant on river navigations in Yorkshire before setting up in business on his own in 1772, having been elected a member of the Society of Civil Engineers in 1771. He went on to acquire considerable experience on river navigations and canals in England and Ireland, developing close contacts with engineering contractors such as the Pinkertons, and Smeaton's withdrawal into semi-retirement left Jessop the senior civil engineer at the beginning of the Canal Mania, at least in so far as water and earth works were concerned. In these years he assumed responsibility for ten canals: Cromford (1789), Leicester (1791), Nottingham (1792), Horncastle (1792), Basingstoke (1793), Leicestershire and Northants (1793), Barnsley (1793), the Grand Junction (1793), the Ellesmere (1793), and the Rochdale (1794) - the last of these having usually been credited to Rennie before the researches of Hadfield and Skempton.[16] On any score, this constitutes an impressive record of engineering activity, especially as Jessop saw all these canal undertakings through to completion.

John Rennie (1761-1821) was sixteen years younger than Jessop and four years younger than Telford, but he rose to prominence as an engineer more rapidly than the latter. His comparatively early rise to fame was made possible partly by his innate ability both in practical and theoretical affairs, and partly by the fact that his family were well-to-do farmers in the fertile plain of Midlothian, so that they could afford to encourage John's practical talents in the neighbouring workshop of Andrew Meikle, and to ensure him an excellent education at the local grammar school and Edinburgh University. In a sense, John Rennie confounds the popular generalisation that the typical British engineer of the Industrial Revolution was a practical man with little, if any, theoretical background.[17] However, the fact that he was exceptional in his grasp of engineering theory, and that, despite his active support for the Society of Civil Engineers, nothing was done during his membership to enhance the theoretical competence of his fellow engineers, shows that the generalisation remains substantially sound. Not for another hundred years after the beginning of Rennie's career could a theoretical education begin to be regarded as important to a British engineer, and even then it remained dispensable.

The coming of the Canal Mania found Rennie in London, where he had moved to execute his first large commission as a millwright in constructing the mill work of Albion Mill in Southwark for Boulton and Watt. He was thus well placed to become heavily involved in canal construction in the 1790s, and he became responsible for the Worcestershire and Birmingham (1791), the Lancaster (1792), and the Kennet and Avon (1794), as well as numerous other ventures which failed to materialise for one reason or other. Perhaps it is appropriate, considering the fame achieved by Rennie as a bridge-builder for his three great bridges over the Thames in London, that the best-known features of these canals should be the monumental but graceful masonry aqueducts over the Lune on the Lancaster and over the Bristol Avon by the Dundas and Avoncliff aqueducts near Bath on the Kennet and Avon. But it is probably true to say that Rennie made his most significant contribution to canal engineering through his organisational skill in devising a managerial structure for large-scale engineering works in which the various functions of consultant engineer, resident engineer, and engineering contractor, were fully and clearly defined. Amongst all the eighteenth century canals, the Kennet and Avon Canal Company best exemplified this structural differentiation. It was carefully worked out by Rennie and provided a model for many subsequent large engineering operations.[18]

Thomas Telford (1757-1834) only began his career as a canal engineer in the Mania years, by working with Jessop on the Ellesmere (1793) and by constructing the Shrewsbury (1793), and the most impressive works on these canals like the great aqueducts were completed some time after the Canal Mania. But he deserves to be included as a member of the triumvirate of talent of these years because from these beginnings he went on to become the outstanding canal engineer of the subsequent period. By his monumental masonry work, and his large-scale introduction of cast iron into canal engineering, particularly in his strikingly novel aqueduct designs, Telford established himself as the last great master of British canal construction.

Other figures of the second period of British canal construction included some significant survivors from the Brindley era, and several men who established their reputations as engineers in fields other than canal work. Robert Mylne, the engineer for the Gloucestershire and Berkeley (1793) was an irascible Scottish architect who achieved fame as the builder of Blackfriars Bridge in London between 1760 and 1769. He played a prominent part in the reorganisation of the Society of Civil Engineers in 1793, although his professional contacts did not prevent him from being abusive about colleagues who questioned his engineering judgment.[19] Like other early civil engineers, Mylne established a dynasty of family members and apprentices who carried on his business after his death. John Watte, the land-drainage engineer responsible for the Wisbech Canal (1794) was another member of the Society. Benjamin Outram, engineer for the Peak Forest (1794) and Huddersfield (1794) was associated with Jessop in the Butterley Company which they established in Derbyshire to provide foundry and machine work, and went on to make a name for himself as a pioneer of tramway construction. The Peak Forest Canal, like several other lines, was deliberately designed with vital sections of tramway to feed it.[20] William Smith, the land surveyor who was commissioned to engineer the Somerset Coal Canal (1794), became involved in an expensive experiment to construct 'caisson locks' in order to conserve water on the change of levels in the canal at Combe Hay, but was dismissed on the failure of this venture. Nevertheless, Smith went on to win a national reputation as the 'Father of English Geology' for his brilliant work on stratigraphy which began through observing fossils on the alignment of the Somerset Coal Canal.[21] William Chapman, also, deserves to be mentioned for his work on the Grand Canal in Ireland. Chapman was a Yorkshireman who was encouraged by Smeaton to enter the engineering profession, and who developed notable skills in canal and bridge construction.[22]

The Canal Mania collapsed in 1795 with the onset of the French Revolutionary Wars. A period of rising prices and prolonged inflation made canal speculations less attractive, and they never regained the brief popularity of the Mania years. Some of the canals promoted in the years of euphoria were abandoned, and most of them took many years to complete as the building operations were slowed down by the lack of adequate finance. Canal work thus continued to account for a considerable amount of civil engineering employment from the mid-1790s until the rapid development of the railways in the 1830s. This was the third period of British canal construction, the generation of the 'tail-enders'. The triumvirate of the Mania years continued to be deeply involved in canal work, with Jessop undertaking the Grand Surrey (1801) and Rennie the Croydon (1801), the Royal Military (1804), the Portsmouth and Arundel (1817), and, in Scotland, the Aberdeen (1796). But Jessop died in 1814 and Rennie in 1821, leaving Telford as the undisputed doyen of latter-day British canal engineering, right into the Railway Age which was beginning when he died in 1834. In these years Telford made significant improvements in the Birmingham canal system, and engineered new canals in the Birmingham and Liverpool Junction (1826) and the Macclesfield (1826) which, with their cuttings, embankments, masonry and cast-iron work represented the mature self-confidence of this last phase of British canal engineering. Telford's most spectacular canal achievements of these years, however, were his two sea-to-sea canals, in Scotland and Sweden. The Caledonian (1803) was part of his great work of opening up the Scottish Highlands by the construction of roads, bridges, harbours and waterways. By the time it was completed in 1822, the strategic incentives which had inspired its construction had virtually disappeared, and it never showed any likelihood of becoming a commercial success, but it survives to the present day as a splendid piece of canal engineering. Of the sixty miles from Corpach in the west to Clachnaharry in the east, forty miles is through navigable lochs, with only about twenty miles of artificial waterway. This, however, involved some particularly difficult ground for the large engineering works which Telford designed, including twenty-eight locks of unusually massive size for British canals.[23]

Telford was invited to design the Götha Canal in Sweden by Count Baltzar von Platen (1766-1829), a diplomat and statesman of considerable determination who managed to survive big upheavals in the Swedish government during the Napoleonic wars and to win the support of successive monarchs for his scheme to link Göteborg in the west, with its access to the North Sea, to the capital city of Stockholm on the Baltic, by a ship canal. The invitation was sent to Telford on 28th April 1808 and he replied on 2nd June. At the end of July he made his first visit to Sweden to survey the line of the canal, returning home in October. Telford made another two-month visit in the summer of 1813 to review progress on the works, but otherwise his supervision was conducted through an extensive correspondence with von Platen, and through the various agents sent over and entrusted to act on his behalf. Four new towns were projected on the route of the canal, but only one of them was built and it was here, in Motala, that the Canal Company established its headquarters. Here also an iron works was set up to manufacture the lock gates and other equipment for the canal, under a Scotsman, Daniel Fraser, who settled in Motala and is buried in the churchyard there. The overall length of the canal is 347 miles, but the large inland lakes and waterways of Sweden reduce the length of artificial cut to sixty miles, only three times that required on the Caledonian Canal, with sixty-five locks compared with the twenty-eight on the British canal. Another interesting comparison between the Caledonian and the Götha canals is that Telford's initial estimate of costs was £400,000 for both: in the case of the Caledonian this figure was more than doubled to a capital cost of £900,000, while the cost of the Götha was five times greater than the estimate by the time of its completion in 1832.[24]

The construction of the Caledonian and Götha Canals were epic achievements of the Canal Age, and both were master-minded by Thomas Telford. Of the other engineers involved in canal building in this last generation of British canal construction, some represented dynasties already established in canal engineering. These included William Whitworth, engineer of the North Wilts (1813), and son of Robert Whitworth; Josias Jessop, the son of William, who undertook the Wey and Arun Junction (1813); and Sir John Rennie, who was knighted after the death of his father and on the completion of the new London Bridge, and who built the Glastonbury Navigation (1827). William Chapman became involved in English canals with the Sheffield (1815) and the Carlisle (1819). Amongst the remaining names - James Green, Ralph Walker, Ralph Dodd, John Taylor, Benjamin Bevan, James Hollingsworth, James Morgan, John Millington, Thomas

Hughes, George Leather, Henry Buck, Richard Buck, Francis Giles, Robert Coad, Robert Retallick, William Cubitt, Thomas Brown, George Fletcher, Hugh Baird, Stephen Ballard and John Killaly - none require much attention. Cubitt was later to achieve distinction as a railway engineer, and James Green deserves notice as the ingenious and persistent advocate of canal lift-locks in place of conventional pound-locks, seven of which he installed on the Grand Western (1796).[25] In general, however, the big prizes had already been allocated between the earlier canal engineers, and these tail-enders found themselves left with the economically marginal projects, and usually struggling with inadequate funds.

Although the coming of the railways brought the Canal Age to a more abrupt end than might otherwise have been the case, the prime cause for the tailing off of civil engineering activity on the canals was the fact that, by 1830, they had virtually all been built and had already provided that significant though limited stimulus to the national economy of which they were capable. As the canals of the Mania years were completed - often long after the year in which they had been started, at much greater cost than had been projected, and without always fulfilling the promise with which they had been welcomed - there was not much left to be done except to make a few links of comparatively marginal economic value, or to undertake improvements to existing canals. The latter course was not regarded with enthusiasm by investors who had frequently had to wait for many years for any returns on their capital, and who were consequently inclined to neglect canal maintenance rather than contemplate new works which would inevitably be disruptive and expensive. Improvements were made, such as the shortening of the Oxford and other early 'contour' canals, and the construction of the Birmingham and Liverpool Junction (1826), Telford's last major canal work. But too little was done too late to make the canals a really viable competitor with the railways, if this could indeed ever have been achieved within the limits of British topography. British canal building, after all, had begun to decline two decades before the coming of the railways. From 1800 onwards it had entered upon a long drawn out twilight in which projects were completed or abandoned, and engineers moved into other areas. It is significant in this respect that Jessop, Rennie and Telford all became heavily involved in dock construction in London and elsewhere as canal construction languished, and went on to work on bridges, waterworks, lighthouses, and land-drainage schemes before the railways began to absorb the attention of their successors.

While creating the first large-scale employment opportunities for civil engineers, the canals also provided a vital stimulus to the formation of the first professional associations of engineers. We have already observed the dominant role of canal engineering in the early years of the Society of Civil Engineers, and this preponderance remained strong after the reorganisation of 1793. It is only possible to outline the activities of the Society in these years, because the sketchy Minute Books tell us little about the business performed. One fact which does emerge strongly, however, is that the Society remained a very small and select body of men. Whatever the causes of the reorganisation in 1793 - and they remain obscure - it is apparent that the membership started with a nucleus of four - Jessop, Rennie, Whitworth and Mylne - and that, after four seasons of meetings in 1796 it was still under thirty.[26] At least, however, the Society was specific about its objectives..Volume II of the 'Treasurer's Minutes and Accounts' runs from 1793 to 1821. The 'Heads of Resolutions for Establishing a Society of Civil Engineers' were set out at the beginning, on 15 April 1793. No one was to be a member 'except those who are actually employed in designing and forming Works of different kinds, in the various departments of Engineering', but provision was made in addition for a class of Gentlemen as honorary members, to include 'various Artists, whose professions and employments are necessary and useful thereto as well as connected with Civil Engineering'. Five members were to form a quorum, and members were to be elected by ballot from candidates proposed and seconded by two members at a previous meeting: two thirds of those voting were to be in favour for a successful application. It was resolved: 'That a Civil Engineer shall preside as Chairman of the meeting for the time being, and that the youngest member present shall act as Secretary'. The annual subscription was to be half a guinea. Members were encouraged to donate copies of their reports and professional papers to the Society. Meetings were to be held fortnightly on Saturdays during the parliamentary sessions.

Despite the quorum rule, few of the early meetings had more than four members present, and on several occasions a single member is recorded as 'solus'.[27] This does not seem to have deterred members from conducting business, including the election of new members, provided that the Treasurer was present, although on some occasions the business was deferred.[28] The first Treasurer was Mylne, and Rennie assumed the duty later, by which time the office had come to include that of Secretary. As far as a Chairman was concerned, the Society stuck diligently to its determination not to have a permanent officer, so that it circulated around members, changing at every meeting. Unfortunately, the only business recorded regularly in the Minutes apart from the election of new members was that concerned with the publication of Smeaton's papers, a long-drawn-out labour of love to which several members, publishers, and booksellers contributed. Otherwise information about what was discussed is sparse: Robert Welldon (sic) 'submitted his invention of a new invented Lock'[29]; Mylne reported that he had obtained for the Society the 'Books and Papers' of the former Society at a cost of half a guinea,[30] and twenty-one years later it was: 'Resolved to buy a large chest' to hold the books and archives of the Society.[31] The Society did not acquire any premises of its own, but the social and convivial aspects of its business are revealed by the regular 'tavern bills' which were paid by the Treasurer for the hospitality of the establishments at which they held their meetings.[32]

As the total number of members remained low it could hardly be expected that regular meetings would be attended by more than a handful, but few attempts were made to boost the numbers of engineers. More attention seems to have been given to recruiting eminent men as honorary members: Sir Joseph Banks was enrolled in 1793, playing an important part in securing the publication of Smeaton's papers; the Earl of Morton in 1796;[33] the Hon. Charles Greville in 1797;[34] Count Romford in 1799;[35] the Astronomer Royal, the Rev. Dr. Maskelyne in 1803;[36] John Hamilton, the Viscount Kirkwall in 1804;[37] William Herschel in 1808;[38] Humphry Davy in 1808;[39] and Davis Giddy in 1811.[40] Amongst the 'artist' or craftsmen members, Ramsden is not recorded as having attended a single meeting, although on his death in 1801, 'Mr. Peter Dolland, of St. Paul's Churchyard' was elected to replace him.[41] But the Society showed remarkably little interest in recruiting new engineering talent. William Mylne was elected to replace his father when he died in 1811, and when: 'It was moved and seconded that the Memory of our late worthy brother Mr. Mylne's name be added to that of Mr. Smeaton in the Toast, on being put to the vote was unanimously carried'.[42] The younger Rennies, George and John, likewise replaced their father when he died.[43] An application from the lighthouse engineer Robert Stevenson was received in 1815 and he was sent a letter 'describing required qualifications', and no more action appears to have been taken.[44] But the biographer of the Stevenson family reports that David Stevenson, one of the next generation of this talented family, did become a member in the 1840s.[45] However, the most glaring omission from any mention amongst lists of members enrolled as attending meetings was that of Thomas Telford, so that it must be assumed that the rivalry and personal hostility often hinted at between Telford and Rennie led to his exclusion from the elite social club in which the latter was one of the dominating personalities.[46] The death of John Rennie was recorded in the Minutes at the beginning of Volume III on 8 March 1822, when the meeting also considered 'the reduced number of their members'. William Chadwell Mylne took on the office of Treasurer and Secretary, and an effort was made to recruit some new engineering members. But by then the response to competition was too late to save the Society as the unchallenged organisation of the profession. The new generation of young engineers had already dismissed it as a vehicle for their professional aspirations and had established the Institution of Civil Engineers with Thomas Telford as their President.

Whatever the merits and achievements of the Society of Civil Engineers, it is certain that the Society had nothing to offer young engineers aspiring to success in their profession. Equally certain, in the decades of commercial disruption which followed the end of the Napoleonic Wars in 1815, and which were characterised by such ugly outbreaks of discontent as the Luddite attacks on machinery and the 'Peterloo Massacre' of 1819, was the fact that the great age of canal building was over, and that the immediate future would be dominated by the steam engine and other mechanical innovations produced in response to advancing industrialisation. What was not so certain was whether or not the rising generation of engineers, experienced with machinery rather than with canals and harbour works, would be able to contrive unaided a new instrument of professional organisation. In the event it transpired that, in the labour of institutional innovation, the canal engineers were able to perform a final outstanding service to their profession.

This institutional innovation was the Institution of Civil Engineers. It was established at a meeting in the Kendal Coffee House, Fleet Street, on 2 January 1818.[47] Eight men came to be regarded as the founding members: William Maudslay, who took the chair at the first meeting, Thomas Maudslay, Joshua Field, Henry Robinson Palmer, Charles Collinge, James Jones, John Letheridge, and James Ashwell. The moving spirit was H. R. Palmer, who appears to have taken the initiative in calling a preliminary meeting at the end of December, and who expounded his view that there was a deplorable lack of professional education for civil engineers and of contact between members of the profession. He characterised the would-be engineer as: 'a mediator between the Philosopher and the working mechanic' (a description frequently echoed in subsequent years as, for example, by Rankine a generation later) and stressed the need of such a professional man for an institution which would enable him to increase his knowledge continuously by contact and discussion with his peers. Palmer was a talented young man who was then an assistant to Thomas Telford, but he had served his apprenticeship with the mechanical engineer Bryan Donkin, and the other founder members were all mechanical engineers.[48] Their youth was indicated by the resolution passed at the first meeting that the age of admission should be between twenty and thirty-five, although this was rescinded in 1820. The other resolutions on this occasion were: 'that a Society be formed consisting of persons studying the profession of a Civil Engineer', that the Society should meet weekly, and that honorary members could be admitted from: 'persons who do not study the profession as a mode of subsistence, but are indefatigable in devoting their leisure to pursuits immediately of that nature'.[49]

Having agreed on the basic principles of the new institution, succeeding meetings hammered out the details of the organisation and also embarked on the earnest discussion of matters of professional interest which became the distinguishing feature of this kind of society. On the organisational side, the rules for elections and for the duties of officers were determined. Field and Palmer accepted the office of Chairman, while Jones became the first Secretary. In 1819, provision was made for the election of 'corresponding members', defined as: 'such persons of the profession as are constantly resident at a distance from London, but are willing to promote the objects of this Institution by communicating such valuable information on professional subjects as they may possess and can make known without inconvenience or injury'.[50] Topics for professional discussion were initially divided into categories: first, 'pieces of direct information in the form of essays etc.', then 'the solution of questions', followed by 'the discussion of the merits of inventions, discoveries and publications',[51] but there is no evidence of these distinctions having been rigorously applied. Discussion ranged over the treatise of R. Buchanan on the shafts of mills,[52] methods of producing parallel motion,[53] and the merits of various forms of canal lock.[54]

Despite the keen support of its early members, the Institution did not show much capacity for growth. The first eight members were joined by only seven in the following two years (Alex William Provis, John Provis, E. A. Morris, and William Warcup on 2 April 1819; John Gibb and Thomas Burnett as Corresponding Members on 11 January and 14 March 1820 respectively - the address of the former was given as 'Aberdeen' and the latter as 'Canada' - and Lt. Gerhard Morris Roentgen on 17 March 1820).[55] The sixteenth member, elected and installed as President on 21 March 1820, was Thomas Telford, and thereafter there was a dramatic acceleration in the recruitment of new members. In this as in other respects, the advent of so distinguished a figure to take the helm transformed a worthy but insignificant club into a dynamic society which quickly asserted and won the position of being the representative organisation of the civil engineering profession.

Telford's election stemmed from a resolution moved by John Provis, seconded by James Jones, and adopted by the meeting on 5 January 1820. The members composed a long letter to Telford, addressed from Gilham's Coffee House on 3 February 1820, and signed by those present. The polite request to the Great Man was couched in some passages of careful self-justification and startling modernity:

'It is unnecessary to remark to you on the business of an engineer; all admit the difficulties of it, and its indefinite character; and that by want of definition, its respectability is less than its due, that public confidence which is indispensable, is much weakened by the presumption of unskilled and illiterate persons taking upon themselves the name. Engineering, indeed, in England, is taught only as a trade, and this is an essential cause of the evil complained of ...

To facilitate the acquirements of knowledge in engineering; to circumscribe the profession; to establish in it the respectability which it merits, and to increase the indispensable public confidence, are the objects of the Institution, the members of which now have the honour of addressing you ...[56]

Telford replied accepting the invitation on 16 March 1820, and gave an inaugural address to the meeting on 21 March 1820.[57] After approving of the steps already taken to establish the Institution, Telford commented:

'In foreign countries similar establishments are instituted by government, and their members and proceedings are under its control; but here, a different course being adopted, it becomes incumbent on each individual member to feel that the very existence and prosperity of the institution depend in no small degree on his personal conduct and exertions; and the merely mentioning the circumstances will, I am convinced, be sufficient to command the best efforts of the present and future members, always keeping in mind that talents and respectability are preferable to numbers, and that from too easy and promiscuous admission, unavoidable, and not infrequently incurable, inconveniences perplex most societies.'[58]

Under Telford's active guidance, there would be no 'promiscuous admission' of new members. Nor, as his biographer observes, was there any danger of the Institution: 'degenerating either into a trades union or a mere debating society'.[59] The genius for organisation which he had demonstrated throughout his career was applied in the 1820s to building up the Institution. He attended meetings regularly, introduced eminent guests, many of whom subsequently became members, encouraged members to offer contributions for discussion, and provided the nucleus for a library from his own collection. The Institution acquired premises of its own in Westminster, and although these changed over the years until the construction of the present palatial offices in Great George Street in 1913, they remained firmly based near to the seat of political and administrative power. In 1825 an amendment to the constitution replaced the previous four 'Chairmen' with four Vice-Presidents, and established a council to direct the business of the Institution, subject only to a general meeting of the members.[60] Further modifications in 1826 defined the grades of membership as ordinary, corresponding, associate, and honorary.[61] At the AGM in 1828, the total number of members was given as 134: thirty-two ordinary, forty-two corresponding, forty-five associates, and fifteen honorary.[62] Attendance at meetings increased gradually, so that more than twenty were regularly present by the end of the decade. However, the Institution at this time was not sufficiently well endowed financially to permit expenditure on anything other than routine activities: a resolution to undertake research in comparing the performance of high and low pressure steam engines was abandoned on account of the 'totally inadequate' funds available.[63] And the annual dinner was discontinued in the same year, ostensibly as being a useless expenditure.[64] Yet the cost of the petition to secure a Charter - some 300 guineas - was raised in 1828 by a loan from two unspecified members.[65]

The attainment of the Charter was a significant mile-stone in the maturation of the Institution, giving a realistic basis to its claim to represent the whole of the engineering profession, even though some of the senior members of the profession, including the second generation of Rennies, withheld their support for several years thereafter: Sir John Rennie was only elected in 1845, immediately prior to becoming the third President. For many years, moreover, the possession of the Charter made the Institution superior to any other engineering association, and the Council tried hard to maintain this distinction by opposing proposals from other bodies to acquire charters for themselves. The Charter also gave the Institution a legal existence, which was important amongst other advantages in that it relieved Telford of the responsibility of acting as its unofficial banker.[66] Of more symbolical significance, the Charter gave wide circulation to the eloquent definition of civil engineering:

'Civil engineering is the art of directing the great sources of power in Nature for the use and convenience of man; being that practical application of the most important principles of natural philosophy which has, in a considerable degree, realised the anticipations of Bacon, and changed the aspect and state of affairs in the whole world ...'[67]

When Thomas Telford died in 1834, the great age of the canal engineers had come to an end, and that of the railway engineers was dawning. Members of the Institution of Civil Engineers had not been unaware of the advent of the railways and steam locomotion, but their contributions remained peripheral to it until the successes of George Stephenson and his North Country colleagues had delineated the future with the success of the Stockton and Darlington Railway in 1825 and, even more impressively, with the opening of the Liverpool and Manchester Railway in 1830. With Telford dead and honoured by entombment in Westminster Abbey, the way became clear for massive involvement by members of the Institution in the new sphere of activity, and as a result it underwent rapid growth and further transformation. The soundness of the essential structure, however, had already been established, and for this Telford deserves much of the credit. True, he was called in to preside over a body which had previously determined its main objectives and principles as an organisation for the mutual edification of engineers and improvement of the engineering profession. But these were objectives of which Telford thoroughly approved, and by enthusiastically placing his great skills and prestige at the disposal of the Institution he enabled it to grow and to begin to realise its potential as the instrument of a united profession in a way which no other initiative could have done. The intense personal effort which Telford put into the organisation is apparent in his correspondence and in the Minutes of the Institution. He missed no opportunity of presenting it to his correspondents at home and abroad, so that the Institution rapidly achieved a significant overseas membership.[68] He rarely missed a meeting when he was in town, usually preceding it with a dinner for guests and resident pupils at his home at 24 Abingdon Street. Several of his pupils spoke warmly of the way in which he had involved them in the early business of the Institution. Joseph Mitchell, for instance, recorded his impressions of being asked to take notes for regular meetings of the Institution in the 1820s, attended by fifteen to twenty members.[69]

The Institution of Civil Engineers was thus a very personal achievement for Thomas Telford, the last of the great canal engineers. When he died in 1834, he bequeathed sums ranging from £200 to £2,000 to over thirty individuals, and £2,000 'To the President for the time being of the Civil Engineer's Institution, in trust, the interest to be expended in annual premiums under the direction of the Council'. He also left 'All my scientific books, bookcases, prints, and such drawings as my Executors shall consider suitable' to the Institution, and this donation formed the nucleus of what quickly became a very substantial library. Telford's will was included as an appendix to the *Life* edited by John Rickman and published in 1838.[70] The volume is based on Telford's own account of his labours, but half the text consists of a selection from his engineering reports, and there is a Preface and Supplement by Rickman.[71] There is little in the seven hundred pages to indicate Telford's commitment to the Institution, apart from the will, a few pages in Rickman's Supplement, and an appendix giving the resolution of the Civils inviting Telford to be their President, and his response to this invitation.[72] Nevertheless, enough is said to confirm the 'intimate connection of Mr Telford with the Institution of Civil Engineers'.[73] It is not exaggerating to claim that in this relationship he fashioned an instrument which proved itself to be capable of great growth and flexibility, and one which, despite the emergence of rival organisations, has never lost its role as the great archetype of engineering professional organisation and the model for engineering societies in Britain and elsewhere.[74]

Notes

1. For this section on canal engineering I have drawn heavily on my article: 'The British Canal Engineers: The Men and their Resources' in Per Sörbom (ed.): *Transport Technology and Social Change*, Tekniska Museet Symposia, Stockholm, 1980, 65-89, and especially the table of 'British canals and their engineers', 81-9.

2. The leading historian on British canals is Charles Hadfield, who has edited the series on 'The Canals of the British Isles' for David and Charles. For an overview of his work, see especially: *British Canals - An Illustrated History*, 1950, 4th edition, David and Charles, Newton Abbot, 1969. Like anyone working on canal history, I am greatly indebted to the work of Mr Hadfield. See also Hugh Malet, *The Canal Duke*, London, 1961.

3. Thomas Telford was largely self-educated and except for his two visits to Sweden never went to continental Europe. Sir Alexander Gibb, however, in *The Story of Telford - The Rise of Civil Engineering*, London, 1935, p. 162, demonstrates that Telford had a keen interest in continental engineering, even though he was critical of it. Dr. S. B. Hamilton tells the story of Telford learning French in order to read Belidor in: 'Continental influences on British civil engineering to 1800' In *Archives Internationales d'Histoire des Sciences*, 2, 1958, 347-55, especially 350.

4. On Liverpool docks engineers in the eighteenth century, see Henry Peet, 'Thomas Steers - The engineer of Liverpool's First Dock - a Memoir', in *Trans. Hist. Soc. Lancs. and Cheshire*, 82, 1930, 163-242; and Stanley A. Harris, 'Henry Berry (1720-1812); Liverpool's Second Dock Engineer', in *Trans. Hist. Soc. Lancs and Cheshire*, 89, 1937, 91-116.

5. The Rideau Canal between Ottawa and Kingston on Lake Ontario was built 1826-32 by the British Royal Engineers. See R. F. Leggett: 'The Jones Falls Dam on the Rideau Canal', *Trans. Newcomen Soc.*, 31, 1957-59, 205-218.

6. The river navigations have not been as well served by historians as the canals. But see T. S. Willans, *River Navigation in England, 1600-1750*, 2nd edition, London 1964; and A. W. Skempton: 'Engineering on the Thames Navigation', *Trans. Newcomen Soc.*, 55, 1983-84, 153-176.

7. See D. H. Tew: 'Canal Lifts And Inclines', *Trans. Newcomen Soc.*, 28, 1951-53, 35-58.

8. The only comparable activity at the time was that of building ships for the Royal Navy, involving the development of the Royal Dockyards. But the men involved in this business could not be regarded as 'civil' engineers.

9. Cyril T. G. Boucher, *James Brindley Engineer 1716-1772*, Goose, Norwich, 1968. The biography of Brindley by S. Smiles in Vol. 1 of *Lives of the Engineers*, Murray, London, 1862, and David and Charles, Newton Abbot, 1968, remains a very entertaining and lively account.

10. This argument is necessarily tentative because of lack of specific indications of Brindley's professional inadequacies. But see Charles Hadfield, 'Rivers and Canals' in A. W. Skempton (ed.), *John Smeaton, FRS*, 1981, for a discussion of disagreements between Brindley and Smeaton on the Forth and Clyde Canal (117-9).

11. Smiles, *op. cit.* David and Charles ed. p. 353.

12. Anthony Burton, *The Canal Builders*, London, 1972, quotes Smeaton (Reports, 28 October 1768) disagreeing with Brindley over the virtue of aqueducts.

13. Foreword to Smiles *op. cit.*, David and Charles ed., volume 2, no page no.

14. Charles Hadfield and A. W. Skempton, *William Jessop, Engineer*, David and Charles, Newton Abbot, 1979.

15. Ibid. p. 261 and p. 40.

16. Ibid. Chapter 6, especially p. 126. Dates in brackets after Canal projects refer to the relevant Acts of Parliament.

17. Smiles *op. cit.* 2 'Life of John Rennie' remains the most accessible account. Cyril T. G. Boucher, *John Rennie 1761-1821 - The Life and Work of a Great Engineer*, Manchester, 1963, has some useful sections on Rennie's theoretical skills (see p. 7, 30-51, etc.)

18. Burton, *op. cit.*, 127-8. Also Kenneth R. Clew, *The Kennet and Avon Canal*, David and Charles, Newton Abbot, p. 51, referring to the K. and A. Western Subcommittee Minute Book, 3 July 1794; and C. T. G. Boucher, Rennie, *op. cit.* 63-65.

19. Burton, *op. cit.*, 106-7, quotes Mylne attributing trouble on his Gloucester and Berkeley Canal: 'From the time of Jessop's visit, I date its misfortune ...' See also A.E. Richardson, *Robert Mylne - Architect and Engineer, 1733 to 1811*, Batsford, London, 1955.

20. See Hadfield and Skempton, *op. cit.*, on Outram. Also Helen Harris, *The Industrial Archaeology of the Peak District*, David and Charles, Newton Abbot, 1971, p. 65 and elsewhere.

21. On William Smith see J. H. Eyles: 'Wm. Smith, some aspects of his life and work', in C. J. Schneer (ed.), *Toward a History of Geology*, Cambridge, Mass., 1969. Also Mrs Eyles' article on Smith in the *Dictionary of Scientific Biography*, New York, 1975, XII.

22. For William Chapman, see A. W. Skempton: 'William Chapman, 1749-1832, Civil Engineer', *Trans. Newcomen Soc.*, 46, 1973-4, 45-82.

23. On the Caledonian Canal, see Smiles *op. cit.* 409-426, and Sir Alexander Gibb, *The Story of Telford - the Rise of Civil Engineering*, London, 1935, 228-9. Also A. D. Cameron, *The Caledonian Canal*, Terence Dalton Ltd., Lavenham, Suffolk, 1972.

24. On the Götha Canal, see Eric de Maré, *Swedish Cross Cut - A Book on the Götha Canal*, Malmo, Sweden, 1964. For the correspondence between Telford and von Platen, see L. T. C. Rolt, *Thomas Telford*, Longman, London, 1958

25. See C. Hadfield: 'James Green as canal engineer', *Jnl. of Transport Hist.* 1, 1953-4, 44-56.

26. The Society of Civil Engineers (SCE) Minute Books are in the Library of the Institution of Civil Engineers. An analysis of the thirty-eight meetings recorded in the years 1793-96 inclusive shows an average attendance of about five: only one meeting reached double figures (11 on 21 March 1794).

27. SCE Minutes: 24 Apl. 1795, 2 Jan. 1807: It was Mylne on both occasions.

28. Ibid., 4 Nov. 1796.

29. Ibid., 30 Jan. 1795: this was Weldon's 'caisson lock' adopted by William Smith for the Somerset Coal Canal, but quickly abandoned.

30. Ibid., 4 Apl. 1794.

31. Ibid., 17 Mar.1815.

32. Ibid., for example 13 Jan. 1797; 23 Feb. 1810; 8 Jun. 1810.

33. Ibid., 11 Mar. 1796.

34. Ibid., 17 Feb. 1797.

35. Ibid., 21 Jun. 1799.

36. Ibid., 20 Jan. 1804.

37. Ibid., 1 Jun. 1804.

38. Ibid., 5 Feb. 1808.

39. Ibid., 20 May 1808.

40. Ibid., 1 Mar. 1811.

41. Ibid., 6 Feb., 20 Feb., and 17 Apl. 1801.

42. Ibid., 17 May and 14 Jun. 1811.

43. Ibid., 26 Mar. 1822.

44. Ibid., 19 May 1815. Stevenson's work on the Bell Rock lighthouse from 1807 to 1810 is vividly described in his own reports as quoted by his grandson, R. L. Stevenson: *Records of a Family of Engineers*, Chatto and Windus, London, 1912. It is significant that this work did not win him access to the SCE.

45. Craig Mair, *A Star for Seamen - The Stevenson Family of Engineers*, Murray, London, 1978, p. 175.

46. Gibb, *op. cit.* p. 191, states that Telford was a member, without providing any documentary basis for this assumption. The Society is not mentioned in Rickman's Life of Telford: see note 70 below. Telford was prepared to speak well of Rennie after the latter's death, but the men seem to have had little time for each other when both were alive.

47. Institution of Civil Engineers (ICE) Muniments: Minute Book 1818-1823 (Reg. no. 93).

48. Joseph Mitchell, *Reminiscences of my Life in the Highlands, 1883*, re-published 1971, David and Charles, in two volumes. Mitchell, recalling the early days of the ICE, described Palmer sharply: 'He was ... a good mechanician and surveyor, self-complacent and indolent. He imagined himself a genius...' (volume 1, p. 91).

49. ICE Minutes, first meeting at Kendal Coffee House, Fleet Street, 2 Jan. 1818.

50. Ibid., 30 Nov. 1819.

51. Ibid., 6 Jan. 1818.

52. Ibid., 17 and 24 Feb. 1818.

53. Ibid., 2 and 24 Mar, 1818.

54. Ibid., 27 Apl., 18 May, and 23 Nov. 1818.

55. List of 'First 100 Members' from F. Graham Clark's mss. notebook in ICE Muniments.

56. Quoted in extenso in Gibb, *op. cit.* p. 194.

57. Gibb, *op. cit.* incorrectly gives 1821.

58. Gibb, *op. cit.* p. 197; ICE Minutes 3 Feb. 1820.

59. Gibb, *op. cit.* p. 199.

60. ICE Minutes 31 Mar. 1825: the first election under this rule was on 19 Apl. 1825.

61. Ibid., 5 May 1826, Special General Meeting.

62. Ibid., 15 Jan. 1828.

63. Ibid., 6 Feb, and 27 Feb. 1827.

64. ICE Minutes of Council 1824-27.

65. ICE Minutes, AGM 15 Jan. 1828.

66. E. Graham Clark: typescript notes for a history of the ICE, p. 23, in ICE Muniments, Reg. no. 282.

67. Quoted L. St. L. Pendred, 'The Institution of Civil Engineers', *British Engineering Societies*, Longman, London, 1947, 14-15. These famous and oft-quoted words were written by Thomas Tredgold (1788-1829), a self taught expert in many branches of engineering and the author of several widely-read technical works: see L. G. Booth, et al: 'Thomas Tredgold (1788-1829) - Some Aspects of his work', *Trans. Newcomen Soc.*, 51, 1979-80, 57-94.

68. His correspondence with the Swedish Baron von Platen illustrates this point. In March 1822, Telford wrote to von Platen, reporting on the progress of the Institution and commenting in particular on its international membership: see L. T. C. Rolt, *Thomas Telford*, 1958, Pelican ed. 1979, p. 202.

69. Mitchell, *op. cit.* 1, 99-100.

70. John Rickman (ed.), *Life of Thomas Telford, Civil Engineer*, London, 1838. There is also a supplementary volume, 'A Folio of Copper Plates'.

71. John Rickman (1771-1840) was a clerk at the House of Commons, and was one of the Executors appointed by Telford. He was also a recipient of one of Telford's donations of £500. He befriended Telford through acting as secretary to the Commissions for making roads, bridges and the Caledonian Canal in Scotland, but the two men also shared literary friendships with Southey and others. Rickman was thus a firm 'establishment' figure, although he had a reputation for sartorial eccentricity, and his main claim to fame was as the virtual creator of the British decennial population census, held first in 1801.

72. Rickman, *op. cit.*, 277-279, and Appendix T, 653-654.

73. Ibid., p. 277.

74. It is of interest that Telford provided in his will that the whole benefaction of books and papers to the Institution should pass to the Royal Society of Edinburgh in the event of 'the said Institution being discontinued': Ibid. p. 662.

Chapter Four

The Beginnings of Institutional Proliferation

In early editions of Samuel Smiles' popular *Life of George Stephenson*, Smiles wrote strongly about the ill effects of the Railway Mania on the engineering profession:

'This boundless speculation of course gave abundant employment to the engineers. They were found ready to attach their names to the most daring and foolish projects, - railways through hills, across arms of the sea, over or under great rivers, spanning valleys at great heights or boring their way under the ground, across barren moor, along precipices, over bogs, and through miles of London streets... No scheme was so mad that it did not find an engineer, so called, ready to endorse it, and give it currency... A thousand guineas was the price charged by one gentleman for the use of his name; and fortunate were the solicitors considered who succeeded in bagging an engineer of reputation for their prospectus.'[1]

He went on to censure the 'low tone of morality' in railway working, and particularly on the part of contractors, the worst feature of the system being, he claimed, that 'the principal engineer himself was occasionally interested as a partner, and shared in the profits of the contract'.[2]

When this passage came to the attention of the Council of the Institution of Civil Engineers in 1857, they dispatched their Secretary, Colonel Manby, to express their objection to Smiles, who responded with an assurance that:

'it had been foreign to his views to write anything which could give umbrage to any Member of the Profession...'[3]

and he undertook further to remove the offensive sections from future editions of the book. It is not explained how the author was persuaded to modify his account, but it is certain that the views originally expressed by Smiles were widely current in the middle decades of the nineteenth century. The prodigious expansion of railway engineering at that time created a demand for engineers which was far in excess of the traditional training procedures to provide recruits, despite the extremely tenuous nature of those procedures. As a result, men of little knowledge or experience were able too frequently to present themselves as 'engineers' to clients who were not too fussy about formal credentials.[4]

Most of the early railway engineers became members of the Institution of Civil Engineers, so that the Institution grew rapidly in numbers during the years of rapid railway expansion, from 288 in 1836 to 797 in 1856. But it did not transform itself sufficiently to retain the wholehearted allegiance of the railway engineers, who in 1847 precipitated the process of institutional proliferation which was to have such serious consequences for the engineering profession in Britain. In this chapter we will be concerned, first, with the development of the Civils as they entered the Railway Age, and then with an account of the formation of the Institution of Mechanical Engineers and its immediate consequences.

At the time of Telford's death in 1834, the Institution of Civil Engineers was facing the dawn of the Railway Age with considerable confidence. Telford himself had had little to do with the railways, but several members of the Institution such as William Cubitt and C. B. Vignoles had already, by the early 1830s, made important contributions to railway works, and a galaxy of talent amongst younger members, which included I. K. Brunel and Robert Stephenson, was ready to shine in the decades of railway expansion which were just beginning. As the Society of Civil Engineers demonstrated no desire to compete with it by recruiting railway engineers, and was content to become an exclusive dining club of senior men in the profession - it even donated its library to the Civils in 1846[5] - the Institution of Civil Engineers was the only professional organisation in the field, and its Royal Charter gave it the authority to speak for the whole body of engineers. Increasing numbers had already put the finances on a sound foundation, so that the Institution possessed its own premises, its own library, enriched with the volumes bequeathed by Telford, and its distinctive system of classified membership. Admittedly, in James Walker, its second President, it had a conventional figure who represented the era of canal construction rather than railway engineering, but he was responsive to many of the pressures for new developments. Under Walker, the Institution acquired its first full-time officer, its first systematic publications, and improvements in its classification of members.

The office of Secretary had been an honorary one before 1839. It had been filled first by James Jones, and then by Arthur Aikin, a founder of the Geological Society and Secretary of the Society of Arts. Subsequently William Gittins became Secretary, and then Thomas Webster, a lawyer who held the office from 1836 to 1839.[6] In the session 1839-40, however, Council decided that the Institution needed the attention of a full-time Secretary and appointed Mr. (subsequently Colonel) Charles Manby in this capacity at midsummer 1839.[7] Manby held the post for seventeen years, and he appears to have acted vigorously in building up the reputation of the Institution. He resigned in 1856, but retained a close association with the Institution as its Honorary Secretary until his death in 1884.[8] The new Acting Secretary in 1856 was James Forrest, who was appointed as Assistant Secretary but became Secretary in 1860 and continued to hold the office until 1896, passing virtually the whole of his career in the service of the Institution. The third full-time Secretary was Dr. J. H. T. Tudsbery, who was still in office in 1914.

One of Manby's first responsibilities as full-time Secretary was to promote the systematic publication of the business of the Institution. The first volume of the *Transactions* had already been issued in 1836, in a handsome quarto format with 325 pages and twenty-eight well-illustrated papers, and two further volumes followed in 1838 and 1842. Meanwhile, in 1837, a smaller publication in octavo size had appeared as *Minutes of Proceedings*, and from 1841 Manby succeeded in producing a combined form of *Proceedings* containing both a record of the annual business of the Institution and a summary, often quite substantial, of the papers delivered during the session. This began a series which has continued ever since. It took the Institution of Civil Engineers twenty-three years to create the resources and to devise this publication, but it immediately became one of the distinguishing marks of any organisation seeking to emulate the Civils that they should possess a similar series, so that the new foundations regarded the publication of their own proceedings in annual volumes on this model as a high priority, and they usually achieved it from their first year.[9]

The first volume of the *Proceedings* covered the sessions from 1837 to 1841. In 1837 the total membership was calculated at 252, consisting of forty-seven Ordinary Members, ninety-three Corresponding Members, ninety-eight Associates and fourteen Honorary Members. The total expenditure for the year was £851.11s.4d. In this year the Institution received a deputation from University College London, asking Council to initiate a class for engineers, and it would appear that the profession was applying itself seriously to the problem of improving the educational quality of recruits.[10] Although nothing came immediately of the approach from UCL, it probably stimulated the modification of the bye-laws in the following year whereby a new class of Graduate Members was created.[11] This change was justified by H. R. Palmer by quoting from his inaugural address of 1818 and by claiming that 'it may with confidence be asserted that this Institution is itself a School of Engineers'.[12] At the same time, the grade of Corresponding Member, which had been intended for engineers not normally resident in London ('beyond the limits of the threepenny post') was merged into the class of Ordinary Members, although a concession was made in the subscription rates for non-resident members.[13]

The Council of the Institution had already emerged as the dominant body in its constitution by the time of Telford's death in 1834, and the *Minutes of Council* thus contain a lively review of the main items of business and controversy undertaken in the middle decades of the nineteenth century. Membership of Council was small and attendance tended to be rather erratic, because the members elected to Council were frequently amongst the most busy engineers of the day. In December 1838 the President was asked to write to Council members W. A. Provis, I. K. Brunel and R. Stephenson asking: 'whether they would be able to give an occasional attendance at the Council during the ensuing session' and it was resolved to insist on the importance of attendance.[14] Brunel replied that as he expected to be less out of town in the future he would be happy to be of service to the Council, but in practice his attendance did not improve significantly.[15] Council also expressed anxiety about members who got into arrears with subscriptions, being reluctant to exercise the 'unpleasant duty' of erasing their names from the register:

'But the consequences of the suspension of the name of a respectable individual or his expulsion from the Institution are so very disagreeable that the Council are unwilling to resort to such a step.'[16]

A more serious constitutional issue arose in the 1840s - the question of the tenure of the office of President. Telford had been automatically re-elected each year during his life, and the practice had been continued with his successor, James Walker, after 1834. Walker seems to have been a respected and competent chairman of Council. At a special meeting held in his absence in 1841, Palmer took the chair and it was resolved to recognise Walker's 'unwearied exertions' on behalf of the Institution by inviting members to make a subscription towards having his portrait painted.[17] However, within a few months there were signs of unrest amongst some sections of the membership. Council Minutes for the autumn of 1841 contain an exchange of letters between the Secretary and Alexander Gordon, a member with a reputation for irascibility,[18] who complained that 'under the close and objectionable system of nomination' for office on the Council, a proposal to nominate him (Gordon) had been blocked by 'a certain party'.[19] When asked to provide evidence Gordon widened the criticism to observe: 'how many senior and distinguished members of the Profession have withdrawn their attendance from the Public Meetings', and to attribute this evil to 'an almost permanent president' when other members were well qualified to fill what he described as this *'highly profitable'* office.[20]

Despite this innuendo of financial impropriety - probably exaggerated, although equally probably not without foundation - Council felt able to ignore Gordon's complaints in 1841, but the point of contention did not disappear entirely and in 1844 it came to the surface again. Discussing the ballot for officers in the following year: 'It was represented that there was reason to believe a feeling existed, that in the present state of the Institution, a shorter period of tenure of the offices of the President and the other members of the Council would be desirable'.[21] In the absence of Walker, the Vice President Joshua Field was deputed to put this suggestion to him, and at the next meeting a letter from Walker was read withdrawing his name from the candidature for President. Council regretted losing: 'so efficient and valuable a President... to whom the best thanks of the Society were justly due', and unanimously decided to invite Sir John Rennie to stand for President.[22] Rennie was duly elected, and served for three years, after which a two-year term became the general pattern until 1880, when W. H. Barlow declined to serve for a second year. Since then, the practice has been established of electing a new President annually.

Whatever the groundswell of feeling amongst the membership, it seems probable that Manby played a large part in this reform of the office of the presidency.[23] The reform resulted in an increased flexibility and a sense of ordered progression amongst senior members of the profession which has been a stabilising factor, ensuring access to new ideas and preventing the accumulation of complaints amongst those members who had previously come to think that the officers were immune to influence from the rank and file. The creation of a strictly rotational system of preferment amongst the elected officers was thus an important reform, completing the institutional model which was then followed by most of the new engineering organisations which followed after 1847.

The maturation of the Council of the Institution of Civil Engineers is apparent in the growing complexity of the business which it conducted in these years. Reference has already been made to the effective protest which it presented to Samuel Smiles in 1857 about passages in his *Life of George Stephenson*,[24] but this was only one of several occasions on which the Council was moved to represent and defend the honour of the profession. It was particularly sensitive to any suspicion of sharp practice, as had been hinted at in Smiles' original text. The charge came up again in 1863 when it was stated in the course of an arbitration procedure that it was customary in the engineering profession to receive commission: 'both from the employers and from the Contractors'. The Council, outraged, gave Manby (by this time he was the Honorary Secretary) instructions to demonstrate: 'that such an assertion was utterly untrue: that the Council could not believe any respectable engineer had acted in the manner described', and that, moreover, if any member of the Institution was proved to have acted in this way he would be expelled.[25] Six years later, Council was moved to issue a similar protest when it received notice of a circular from the Government of India containing a clear understanding that it was the recognised practice in Britain for civil engineers to get a commission on contracts given and materials ordered. Council responded indignantly that: 'any Civil Engineer detected in such practices would be held by the profession to be guilty of disgraceful conduct, which would disqualify him from being a member of this Institution'. The Government of India was censured for having published such imputations without investigation, and it was asked to withdraw the statement.[26] The Secretary of State for India, the Duke of Argyll, was approached[27] and an apology was eventually received from the Governor General of India.[28] Whether or not the misunderstanding arose from a reading of Smiles, or from a tradition of laxity in the early days of railway speculation which had presumably been Smiles' source, it is impossible to say. But the Council undoubtedly enhanced the credit of the profession by responding so firmly whenever such damaging imputations received any sort of official encouragement.

Such firmness required some disciplinary strength to he completely convincing, and here the Council found itself in rather a quandary. Apart from dropping members off the register for non-payment of subscriptions, there is very little evidence of disciplinary activity on the part of the Council in these years, and it appears that the Institution had doubts about its powers over members. Legal advice was sought in 1867, when it was determined that Council could indeed expel a member for misconduct, but not: 'at the arbitrary will of the Council or a majority of the members'.[29] Nevertheless, it was deemed necessary to modify the bye-laws in order to remove any possibility of doubt about the powers of Council in this matter. The question had been posed in an exceptionally delicate form by the censure of a Vice-President, John Scott Russell, earlier in the same year. Russell had been charged with unprofessional conduct involving financial improprieties in the course of acting as agent for various American interests during the Civil War. It was a complicated business, and Russell's modern biographer has gone to some pains to exonerate Russell from the charges which he claims were wrongly brought against him as the result of a vindictive feud.[30] There is some substance in this defence, but Russell, despite his unfailingly suave manner - which seems itself to have been a subject of irritation - cannot escape unimpaired from the charges of financial mismanagement and political duplicity. Council, which held long sessions to resolve this problem, decided that it was not an expulsion issue, although Russell was required to resign from the Council and thus forfeited his opportunity of becoming President.

Although the matters dealt with by Council were usually constitutional and business affairs, they had also to deal on occasion with reactions to the main activities of the Institution - the programme of weekly meetings through which, as a result of the regular publication of its proceedings, the members of the Institution continued to conduct an impressive amount of self-education. The papers presented and discussed, week by week, comprised the core-function of the Institution, and there can be no doubt that they maintained a steady production of high-quality technical information on matters of great current interest to the professional engineers of the day. The discussion of these papers often involved serious differences of opinion, and occasionally generated bad-feeling or misunderstanding which required the intervention of Council. One such case occurred in 1852, when Captain Simmons, R.E., a member of the Institution and one of the Government Inspectors of Railways, complained about the report of a discussion in the *Proceedings*. This concerned a delay of four months which he, as an Inspector, had placed upon the opening of the Torksey Bridge over the River Trent, pending further tests and modifications.[31] The issue was a sensitive one because the profession had responded critically to the decision of the Government to conduct an enquiry into the strength of railway bridges after the collapse of Robert Stephenson's Dee Bridge in 1847. The railway engineers feared that intervention by Government would restrict their own attempts to design stronger bridges and make the problem worse rather than better. Simmons' decision to delay the opening of the wrought-iron girder bridge designed by John Fowler, one of the new young railway engineers, sent a frisson of anxiety through the profession, so that the leading participants in the case were invited to speak at a meeting of the Institution. The discussion went over several meetings, and Simmons was probably correct in complaining that the report in the *Proceedings* represented an expression of official opinion by the Council which he considered to be unjustified. But Council disagreed:

'... there is no doubt, that the summary does convey a correct exposition of the general and almost unanimous sentiments of the speakers, and also that the principal members of the profession coincide in the opinion that ... commercial enterprise is liable to be seriously prejudiced by the system of Government supervision which is operating so injuriously abroad and which has recently been introduced into this country.'[32]

Simmons retorted that the Institution should not express authoritative opinions on scientific and other matters brought before it, but Council refused to withdraw, and after further exchanges the correspondence was formally closed.[33]

The case of the Torksey Bridge nicely illustrates the cautious but effective manner in which Council occasionally expressed a firm collective opinion and thus began to acquire a role as the mouthpiece for the profession on matters of serious concern. Similar issues in the middle decades of the nineteenth century included the promotion of research into atmospheric railways, provided: 'that the investigation does not extend to the commercial value of the Atmospheric system, but is strictly limited to the proposed scientific object';[34] support for the Royal Commission organising the exhibition of Industry of All Nations in 1851;[35] help in finding engineers suitable for work in the Crimea in support of the armed services;[36] and opposition to the plan to establish an engineering college at Cooper's Hill to train engineers for government service in India, on the grounds that it: 'implies that the present means of attaining an adequate Engineering training in this country are defective'.[37]

The last of these points raises a question which frequently came to the attention of the Council - that of the education and training of engineers. We have already noted an early approach from University College, London, inviting the co-operation of the Institution in setting up a class of students, and this was followed up some years later by a request for help in examining undergraduate students. The Institution responded cautiously and inconclusively.[38] The general attitude of the profession in these years was conservative on educational policy: engineers considered that the traditional system of pupilage had served them well, and they were reluctant to consider innovations. But the Presidential Address of John Fowler in 1866 touched a nerve of uneasiness about the emerging inadequacies of this system, and stimulated a proposal for a 'junior engineering society' in order: 'to enable the rising generation of Civil Engineers worthily to maintain the reputation of their predecessors in the face of the strenuous and increasing competition of their rivals on the Continent and in America'.[39] Council responded sympathetically, but eventually decided to contain the proposed development within its own framework by establishing a class of 'Students' who would replace the 'Graduates' and who would be permitted a measure of autonomy in planning their own meetings.[40] This could be seen as the first significant step towards reversing the centrifugal tendency of institutional proliferation, but it did not make much educational impact. When the Institution made its next major public statement on education with the issue of its report on *The education and status of civil engineers* in 1870,[41] it remained strongly traditional and complacent in its emphasis, and it was another quarter of a century before major changes occurred in the attitude of the Institution on this subject.[42]

This discussion of the public role adopted by the Council of the Civils leaves one more question to be resolved: how representative did the Civils see themselves of the profession as a whole? The answer is that, for these middle decades of the nineteenth century, the Institution had no reservations about seeing itself as the general mouthpiece for the profession. The question was posed specifically in 1866 by Henry Cole, who wrote asking if the Institution represented the engineers of the United Kingdom or only England? Fowler replied as President, stating that the Institution:

'is the only chartered Society of Engineers and all the leading members of the profession in Great Britain and Ireland belong to it, as well as nearly all British Engineers of Character and position practising on the continent or in the Colonies.'[43]

He went on to list several 'subordinate and local societies' including the Institution of Mechanical Engineers at Birmingham, the Institution of Engineers and Shipbuilders in Scotland at Glasgow, the North of England Institute at Newcastle, and the Institution of Civil Engineers of Ireland at Dublin. But these were, he claimed:

'established for local convenience or as a subdivision of the different branches of the profession, and cannot in any way be considered as "representative". You may therefore safely take for granted that all engineers of eminence or position whether in England Ireland Wales Scotland or the Colonies recognise the Chartered 'Institution of Civil Engineers' as their sole representative body.'[44]

This very important statement explains why the Civils felt so strongly when other engineering societies sought charters for themselves, because any such recognition of the corporate character of other institutions was tantamount to admitting that they had rivals as representatives of the profession. Thus, in 1867 the Institution petitioned the Queen to request that a charter of incorporation should not be granted to the 'Society of Engineers', a body founded in 1854 'for the discussion of scientific and other subjects of general interest'. The petition set out the grounds on which the charter of the Institution had been granted in 1828, since when it had been recognised: 'by the British Government... and also by Foreign Powers as the representative Body of the Engineering Profession in the

United Kingdom'.[45] A similar statement was made in 1880 in the form of a letter to the Privy Council, objecting to the granting of a charter to the Society of Telegraph Engineers.[46] Eventually, the Civils were obliged to recognise that rival bodies had emerged in fact and that to object to the corporate name was futile once the reality had been accepted. But up until the 1880s they continued to regard themselves as the sole representative of the profession, and to act as such by offering help and copies of literature to bodies elsewhere: to Dutch engineers in 1848,[47] to engineers in New South Wales and Egypt in 1857,[48] to the Institution of Mechanical Engineers in 1857,[49] to engineers at the Cape of Good Hope in 1858,[50] to St. Petersburg engineers in 1861,[51] to Melbourne engineers in 1865,[52] to engineers in Glasgow in 1867 and in Edinburgh in 1869,[53] and to the American Society of Civil Engineers in 1870.[54] These links were frequently only presentations of publications, but they represented a genuine exercise of helpful leadership by a representative parent-society towards other organisations for which it felt some responsibility.

Even though the Institution of Civil Engineers seemed to exercise wide influence and authority over the engineering profession in the middle decades of the nineteenth century, it was in these years that its claim to represent the whole of the profession was first challenged. At first the challenge was muted and even somewhat apologetic, but it gathered force with the establishment of other organisations which claimed to speak for particular local or specialist groups of engineers, so that by the 1880s the Civils had been forced to recognise the diminution in their sphere of influence. As the profession suffered from this self-inflicted fragmentation, it raises the important historical question: why did it happen? In order to answer this question, it is necessary to recognise the pressure generated within the profession by different types of engineering. The British railways were built almost entirely under the direction of civil engineers, but in the course of railway construction and in the subsequent growth in the need to equip and maintain the railways, a new type of professional engineer became prominent - the mechanical engineer. Strictly speaking, of course, the type was not new, because it derived from the same mechanics, millwrights, locksmiths, instruments makers, and so on, who had been amongst the pioneers of civil engineering, and whose numbers included James Watt, Joseph Bramah, Henry Maudslay, Joshua Field, and William Fairbairn - some of the most distinguished members of the engineering profession. As we have seen, the initial group of young men who had established the Institution of Civil Engineers had been predominantly men of mechanical background and training rather than canal or harbour engineers, but the fact that such men were content to coexist with more 'traditional' civil engineers in the same Institution gave that body a cohesion and unanimity which served the profession well for three decades.

It was the unprecedented growth of the railways into large-scale undertakings requiring a permanent staff of engineers which upset this harmonious balance, and promoted the formation of a specialised institution to cater for their type of mechanical engineering. This became apparent after 1847, when the example of the Institution of Mechanical Engineers provided a model for many other regional and specialist associations. Nevertheless, the railways were built largely by civil engineers. Apart from George Stephenson, whose case was so exceptional that we will return to it shortly, the bulk of the early railways were built by the great 'Railway Triumvirate' of Robert Stephenson, I. K. Brunel, and Joseph Locke. Robert Stephenson (1803-59) was born with the outstanding advantage of being the son of the man who was the genius of British railways and the prophet who accurately foresaw the realisation of a national network of rail transport. But Robert added to his father's qualities a sensitivity and talent which made him an exceptional innovator in the design of locomotives and iron bridges, and a social charm which made him acceptable in circles in which his father was never comfortable. He was childless and suffered a great personal tragedy in the loss of his wife in 1842. Like all the leading engineers of the period, he worked with great application and intensity for long periods, although he eased off his professional commitments towards the end of his life, delegating work with increasing confidence to his able staff and seeking solace in sailing his yacht *Titania*. His principal railway works were the London and Birmingham, the Newcastle and Darlington Junction, and the Chester and Holyhead. He also built several lines overseas. Like his peers, Stephenson was a loyal member of the Civils. He joined in 1830, became a Council member for several years, and eventually served as Vice-President and then as President in 1856-57 although this did not prevent him from joining the Mechanicals and succeeding his father as the second President of that Institution in 1849. When he died in October 1859 he left a considerable fortune, including legacies of £2,000 to both Institutions.[55]

I. K. Brunel (1806-1859) was three years younger than Stephenson, but died a month before him in September 1859. Into his short life of fifty-three years he packed an enormous amount of engineering creativity. Like Stephenson, he had the advantage of a father who was an accomplished engineer, but Marc Isambard Brunel (1769-1849), French emigré and builder of the first tunnel under the Thames, was not a railway man like George Stephenson, so that his son was not specifically trained for a railway career. That, however, was no disadvantage, because Isambard Kingdom Brunel was an engineering polymath. Not only did he engineer the superlative Great Western Railway system, with its easy gradients and broad gauge track, but he also built splendid bridges at Clifton, Chepstow, and Saltash, and established himself as the outstanding shipbuilder of his age. Brunel had a very strong professional consciousness, so that he took any opportunity offered him to expound the qualities required of a professional engineer. He was a keen supporter of the Civils, even though his attendance at meetings was erratic on account of his massive business commitments, and he certainly seems to have taken his membership of this Institution more seriously than that of the Royal Society or of his other affiliations. Elected in 1830, he was a Council member in the 1830s and became a Vice-President, but the combination of pressing business undertakings and the onset of mortal illness prevented him from following Stephenson as President of the Institution.[56]

Brunel's place as successor to Stephenson in the presidential chair of the Institution of Civil Engineers was taken by Joseph Locke (1805-1860), who with Stephenson and Brunel made up the 'Railway Triumvirate' of the great age of British railway construction. Unlike the other two, Locke was not the son of a distinguished engineer, but his father worked in the coal-mining industry and knew George Stephenson, who arranged for Joseph to be trained as a railway engineer on favourable terms and then employed him as his assistant on the Liverpool and Manchester Railway. He subsequently quarrelled with the Stephensons, but set out on his own on a brilliantly successful career as a railway builder, being responsible amongst other lines for the Grand Junction and the London and Southampton. Locke's success derived from his skill in keeping his costs within his estimates, so that he acquired the reputation of being an economical engineer. He achieved this objective by avoiding heavy engineering works as much as possible, preferring to give his lines unusually steep gradients rather than undertaking expensive cutting and tunnelling, in the belief, only partially fulfilled, that more powerful locomotives would eventually make these gradients insignificant. Like Stephenson, with whom he was eventually reconciled, and Brunel, Locke was a loyal supporter of the Civils. It was his lot as President to give the eulogies for his two colleagues and rivals when they died in the autumn of 1859, but within a year he also died as a result of an illness which struck suddenly in the course of a grouse-shooting holiday in Moffat.[57]

Between them, Stephenson, Brunel and Locke constructed most of the main line railways in Britain in the middle decades of the nineteenth century, but even in these years there were many miles of track for which other engineers were responsible, and some of these were men of considerable distinction whose contribution might have been more memorable if circumstances had favoured them. William Cubitt, for instance (1785-1861), was responsible for the South East Railway and the Great Northern Railway, and was knighted for his supervision of the Great Exhibition in 1851: he was elected a member of the Civils in 1823 and served as President in 1850-52.[58] Charles Blacker Vignoles (1793-1875) was another railway pioneer, who made surveys for the Liverpool and Manchester Railway and served as Chief Engineer to the Sheffield, Ashton and Manchester Railway, and several other lines in England, Ireland, and the continent: he went off to build bridges and railways in Russia and Spain, but although he worked with the same intensity as so many of his contemporaries, he managed to survive them to reach old age, and he became President of the Civils, which he had joined in 1827, in 1869-70.[59] George Parker Bidder (1806-78), the 'calculating boy' who became a close personal friend of Robert Stephenson and who took a leading part in many of the railway projects of these years, including the Great Eastern Railway, joined the Civils in 1835 and was President in 1860-1.[60] And amongst those engineers who lived to a ripe old age, William Henry Barlow (1812-1902) became the Chief Engineer of the Midland Railway in 1844, joining the Civils in the following year and serving as President in 1879-80.[61]

Apart from these leading figures in the engineering profession, a large number of the less well-known personnel who were assistants and resident engineers to the more prominent men were also keen adherents to the Institution of Civil Engineers. But as the boom in railway engineering developed, an increasing proportion of the whole body of railway engineers did not become members, and amongst the number of those without institutional affiliations was one outstanding figure - the 'Father of Railways' himself, George Stephenson (1781-1848). There has been a protracted controversy about the elder Stephenson's relationship with the Civils, ever since Samuel Smiles gave credence to the legend that he had been denied membership until he had fulfilled the requirements stipulated for new members. These requirements had been defined with increasing precision in the early years of the Institution, but consisted basically of finding three Members prepared to sponsor the candidature and an undertaking to submit a paper or some document of engineering interest to the Institution each year. It seems likely that Smiles had this in mind when he surmised that Stephenson was required to submit: 'a probationary essay as proof of his capacity as an engineer', which he understandably refused to do.[62] It is extremely unlikely that Stephenson ever made an application to join. For one thing, he usually regarded the leading figures in the profession with some disdain because of their opposition to his railway schemes. As Smiles recognised:

'No wonder that he should have been intolerant of that professional gladiatorship against which his life had been one prolonged struggle. Nor could he forget that the engineering class had been arrayed against him during his arduous battle for the Locomotive, and that but for his own pluck and persistency, they could have strangled it in the cradle.'[63]

Amongst the engineers who had incurred Stephenson's antipathy on this score were H. R. Palmer and Thomas Telford, so that it is hardly likely that he could have been well disposed towards the Institution with which they were so intimately associated.

Secondly, it should be observed that no overt hostility towards Stephenson and his family was ever displayed by the Institution. Quite the contrary: Telford himself had introduced the young Robert Stephenson as a member in 1830, and it is inconceivable that he would have done so if he had harboured any antipathy towards the father. Thereafter, the elder Stephenson had a devoted son and other close friends such as G. P. Bidder amongst leading members of the Institution, and again it is inconceivable that, had the old man so desired it, sponsors would not have been forthcoming in abundance. The familiarity of these relationships is exemplified in the delightful anecdote which tells how George used to enjoy wrestling for a throw with Bidder, usually when they met in Robert's office: Robert subsequently sent them bills for damage to his furniture.[64] Again, the warmth of feeling towards George Stephenson amongst all senior engineers flows out of the eulogy made by Joshua Field, speaking as President at the 1849 Annual Meeting of the Civils after George's death, even though he had not been a member.[65] There is thus no evidence that George Stephenson ever applied for membership, and little likelihood of him having done so.

Moreover, there is on record Stephenson's own declaration that he never sought membership of the Civils. In a letter of 27 February 1847 he wrote:

'... I have to state that I have no flourishes to my name, either before it or after, and I think it will be as well if you merely say Geo. Stephenson. It is true that I am a Belgian Knight, but I do not wish to have any use made of it.

I have had the honour of Knighthood of my own country made to me several times, but would not have it; I have been invited to become a Fellow of the Royal Society; and also of the Civil Engineers Society, but I objected to these empty additions to my name. I have however now consented to become President to I believe a highly respectable mechanics' Institute at Birmingham...'[66]

There can be little doubt that the 'Civil Engineers Society' to which Stephenson referred was the Institution of Civil Engineers, and that the 'respectable mechanics' Institute' of the last clause was the new Institution of Mechanical Engineers. As there is no reason to question Stephenson's recollection or veracity on this point, it can be concluded that his stubborn independence caused him to distance himself from the Institution of Civil Engineers right up to his death.

Even though there is now no reason to believe that the engineering profession deliberately spurned George Stephenson, the fact that a large number of railway engineers felt that his role in the railway revolution had not been adequately recognised was significant, because it became an important factor in the creation of the Institution of Mechanical Engineers. Here again the actual circumstances have been obscured by an oft-recounted legend. According to R.H. Parsons, official historian to the Mechanicals:

'The story goes that one afternoon in 1846 a group of railway engineers were watching some locomotive trials on the steep bank known as the Lickey Incline at Blackwell, near Bromsgrove on the Bristol and Birmingham Line. It started to rain heavily and the party took shelter in a platelayer's hut, where the discussion turned on the refusal of the Council of the Institution of Civil Engineers ... to admit George Stephenson...'[67]

Parsons goes on to repeat Smiles' account of the supposed treatment of Stephenson by the Civils. Wherever they met, however, and whatever the particular grievances which they discussed, it is clear that the founders of the new Institution felt deprived of a professional service and were determined to take steps to acquire it, because they issued a circular by way of invitation to fellow engineers to join them in a new enterprise:

'To enable Mechanics and Engineers engaged in the different Manufactories, Railways and other Establishments in the Kingdom to meet and correspond, and by mutual exchange of ideas respecting improvements in the various branches of Mechanical Science, to increase their knowledge and give an impulse to Inventions likely to be useful to the World.'[68]

This was distributed under the unequivocal heading: 'Institution of Mechanical Engineers'. The signatories were James Edward McConnell, the locomotive engineer of the Bristol and Birmingham Railway, to whose house near the Lickey Incline the founders are supposed to have adjourned after their meeting in the platelayer's hut; Archibald Slate, of the Birmingham Patent Tube Co., Smethwick; Charles Frederick Beyer, of Sharp Bros. and Co. Locomotive Works at Manchester; and Edward Humphrys, manager of Messrs. J. & G. Rennie, marine engine builders of Blackfriars, London. Other founding associates were Richard Peacock, Locomotive Superintendent of the Manchester and Sheffield Railway; George Selby, of the Birmingham Patent Tube Co.; and Charles Geach, a Birmingham banker. Apart from Humphrys, therefore, the initial emphasis of the membership was overwhelmingly provincial rather than metropolitan, with Birmingham and Manchester most strongly represented.

It is not known how all the engineers invited to attend the first meeting responded, although it is likely that several replied with caution like I. K. Brunel:

'I beg to record my thanks to Mr. McConnell and the other members of the Committee of the proposed Institution of Mechanical Engineers for the honour they have done me. Will you oblige me by informing me whether the Institution is proposed to be of a local character or as an Institution for England generally as in the latter case I fear it would tend to create a division in our Institution of Engineers and so far would I think be open to objection.'[69]

Brunel did not join the new body, and his fears for the seamless fabric of the engineering profession were eventually fulfilled. But the immediate response to the circular was encouraging, with a good selection of engineers attending the meeting and dinner held at the Queen's Hotel, Birmingham, on Wednesday 2nd October 1846. A committee was formed consisting of Messrs. Peacock, Slate, Humphrys and McConnell, together with William Buckle of Messrs. James Watt and Co.; John Edward Clift, of the Staffordshire and Birmingham Gas Works; and Edward A. Cowper of Messrs. Fox and Henderson, Smethwick - the Midlands bias thus being firmly reinforced. This committee approved rules in November and on 27 January 1847 a general meeting was called in Birmingham for the formal Foundation of the Institution. McConnell took the chair. George Stephenson attended and was elected the first President by acclamation. Justice was thus done at last to the 'Father of Railways', but as on his own account he had become President of what he believed to be 'a highly respectable mechanics' Institute' it is doubtful that he saw any professional honour in the title. Stephenson had been a long-standing supporter of the Mechanics' Institute movement, which made some progress in the 1820s and 1830s as an adult educational opportunity for the artisan and labouring classes, but by the 1840s its appeal had already begun to move up the social scale, so that it would not have appeared incongruous to Stephenson for a 'highly respectable' group of professional engineers to be associated with it.[70] Whatever he understood the new organisation to be, Stephenson took his presidency seriously and presented a paper on 'The Fallacies of the Rotary Engine' in July 1848. He died a few weeks later on 12 August 1848, so he did not live long enough to be disillusioned about the development of the Institution.

Not that disillusionment would have come easily. The process of maturation by which the Institution of Mechanical Engineers was transformed from a provincial organisation of engineers seeking mutual instruction and comradeship into a large national organisation competing with the Institution of Civil Engineers for a significant class of its membership, took place gradually between 1847 and 1877, and it is doubtful whether any except a handful of the leading members were conscious of this trend. First and foremost the Institution satisfied the professional needs of provincial engineers, in ways in which the metropolitan-biased Institution of Civil Engineers could not. Ironically, just as the coming of the railways demolished the need for regional groupings by making it possible for engineers to travel to London comparatively easily, so the increasing number of railway engineers in the provinces made such journeys inconvenient because they had no professional duties in London. Whereas the founders of the Society of Civil Engineers and of the Institution of Civil Engineers had either lived and worked in London, or had cause to visit London regularly in the course of parliamentary and related business, the founders of the Institution of Mechanical Engineers were mainly involved in companies which had little business in London and no requirements for statutory regulations which made representation before parliamentary committees necessary. They were owners, employers, and agents for established railway companies, machine-tool manufacturers, makers of textile and mining machinery, marine engine-builders, and similar occupations. They were concerned with building locomotives and rolling stock, lathes and steam hammers, and steam engines for ships and industrial applications. For them all, a journey to London was either a nuisance or a rare treat, and the opportunity of a professional association meeting at Birmingham was a very welcome alternative.

But even Birmingham did not suit everybody. For the machine manufacturers of Manchester, the mining engineers of Newcastle, and the marine engine-builders of Clydeside - to mention only some of the more important groups of provincial engineers - Birmingham could seem as remote as London, and was certainly not within easy reach for professional or convivial meetings. It is significant, therefore, that while the new Institution found solid support amongst engineers in the Midlands in the years between 1847 and 1855, it was only the introduction of peripatetic summer meetings in 1856 which opened up the attractions of membership to a large provincial catchment area and inaugurated a steady climb in the membership. Unlike the Civils, which had had to struggle to get any members at all in their early years, the Mechanicals had few teething troubles. They recruited seventy members at the outset and grew to 107 in their first year, their meetings frequently being attended by a large number of visiting non-members, including thirty-five at the first AGM. By 1849 they had reached 201 members, but then growth stopped so that the figure had only reached 218 in 1855. Of these, 200 were Ordinary members, two were Life members (Robert Stephenson and H. Maudslay), twelve were Honorary members, and two were Honorary Life members (Eaton Hodgkinson and James McGregor).[71] After 1855, however, the practice of holding a meeting annually: 'in the principal Mechanical Towns of England, Scotland and Ireland, such as Newcastle, Glasgow etc.'[72] led to widening horizons and a resumption in the growth of members, so that the total had reached 1,075 by 1877, when the Institution decided to move its headquarters to London. Paradoxically, the success of the Institution as a provincial organisation promoted the recognition of London as the best national centre from which to administer such an organisation.

By 1877, therefore, Stephenson's conception of a 'respectable mechanics' Institute' had faded away and Brunel's apprehension of a rival to the Civils for the national loyalty of engineers had been partially fulfilled. The element of rivalry, however, remained muted. The Mechanicals had from the outset enjoyed the support of a significant number of members of what was frequently referred to as 'the parent Institution', and the Civils were generally included high amongst the list of toasts at formal dinners of the Mechanicals. In the eyes of many senior civil engineers, moreover, the Mechanicals were conceived as having a function different from and complementary to themselves. This was stated most clearly by Joshua Field, himself a prominent mechanical engineer in the firm of Maudslay Sons and Field in London, when he delivered his eulogy on George Stephenson in the course of his Presidential Address to the Civils in 1849:

> 'George Stephenson was well known to us all. He was President of a kindred society, the Institution of Mechanical Engineers. That Institution principally directs its attention to new inventions in the mechanical and manufacturing processes whilst the communications brought before us (the Civils) are chiefly accounts of executed works in Civil and Mechanical Engineering.'[73]

This was a useful distinction in so far as it helped to define different fields of interest, and the fact that it could not be maintained in practice did not diminish its value as a justification for co-existence. There was thus very little uneasiness expressed by the Civils about what could have been regarded as a cuckoo in the institutional nest. Instead, in 1857 they decided to send copies of their *Proceedings* to the library of the Mechanicals,[74] and over the years many leading engineers held membership of both institutions, and some, beginning with Robert Stephenson who succeeded his father as President of the Mechanicals in 1848-9, became President of both.

The main reason for the low-key rivalry between the Civils and the Mechanicals, however, was the recognition from the outset that they were appealing to different groups of engineers. To this extent, the Mechanicals fulfilled the function of a specialist section, attracting the provincial engineers whom the Civils would not normally expect to reach. A wealth of provincial talent was provided by the railway workshops and by men such as William Fairbairn, Joseph Whitworth, William Armstrong, Robert Napier, and John Ramsbottom, who all became President of the Mechanicals and who were, first and foremost, mechanical engineers, although several of them also rose to eminence in the Civils. The fact that the founders of the Mechanicals modelled their organisation closely on the Civils could thus be accepted as flattering praise rather than as a source of anxiety. Thus, the Mechanicals accepted a similar constitution with similar grades of membership. They started issuing their own publications very promptly and appointed their own full-time paid secretary in 1848 - he was Archibald Kintrea, of whom little seems to be known.[75] He resigned the following year and was replaced by William Prime Marshall, who had served as an assistant engineer to Robert Stephenson and had been Chief Engineer of the Norfolk Railway. The Council was enlarged from five to fifteen members in 1849, with John Ramsbottom, Chief Engineer of the London and North West Railway, becoming one of the leading figures. The Institution acquired its own home in 1850, when it rented a building in Newhall Street, Birmingham. The Secretary lived on the premises, which included a library and meeting room. Meetings were also held occasionally in London, to coincide with the Great Exhibition in 1851, and again in 1852, and there were meetings in Manchester in 1853 and 1855, but it was the decision to hold a Summer Meeting in Glasgow in 1856 which initiated the peripatetic pattern of business already described, and which has served the Institution well.

The development of the Institution of Mechanical Engineers was undoubtedly a success story, but there was an element of ambiguity about the success. On the one hand, the model perfected by the Civils was successfully replicated and applied to a large and important class of provincial engineers who had been virtually untouched by the organisation of the Civils. On the other hand, the development created a second national body which, although it deferred to the Civils in many matters of professional policy, exercised an increasingly independent authority. This independence, backed by growing memberships and influence, served to divide the hitherto unchallenged unity of the profession. Moreover, the very success of the Mechanicals in catering for the needs of one specialised group of engineers encouraged the interest of other specialist groups in seeking independent institutional representation, and the very successful provincial Summer Meetings of the Mechanicals served as a direct stimulus to the formation of new regional associations such as the Institution of Engineers in Scotland. Thus British engineers embarked almost unconsciously on a process of institutional proliferation which, despite all the advantages in terms of numbers and activities, represented an increasing fragmentation of the profession.

Why, then, did it happen? The first cause, as we have seen, was the dramatic rise in the number of engineers brought by the boom in railways, and the second cause was the identity of interest between many members of this new community of railway engineers and those mechanical engineers in the Midlands and the North for whom the Civils were too remote and metropolitan. The desire to honour George Stephenson provided this group with a pretext or justification for establishing a new institution, but once created the momentum generated by an expanding economy in the second half of the nineteenth century, together with a continuing demand for engineering skills, ensured its development into a national body and thus signalled the start of professional fragmentation.

One other cause should be mentioned. In a sense, fragmentation occurred because the Civils made no move to prevent it. The conception of an engineering institution as a medium of collective education seemed to depend upon the preservation of the intimacy of the meetings, and thus dictated a maximum size beyond which an institution could not be expected to grow without losing its club-like character. And without growing in size, it could not possibly cater for the technical interests of all the new specialist groups which emerged in the profession from the 1840s onwards. The price of avoiding fragmentation would have been for the Civils to have recognised the needs of such new specialist skills by establishing sections or groups within its own organisation to cater for them. The creation of a class of Student members with a limited autonomy to arrange their own meetings within the organisation was a step in this direction, but the Civils were unable to envisage an extension of this pattern in the middle of the nineteenth century. Such a solution could have permitted the indefinite expansion of the Institution while encouraging the self-consciousness of its specialist parts and providing ample means for their expression. All the large institutions adopted such a solution by the end of the century in order to accommodate the proliferation of minority interests. But the inability of the Civils to grasp this solution in the 1840s removed the only viable alternative to fragmentation. For a generation thereafter, the Council of the Civils continued positively to welcome the creation of new engineering institutions, as a means of providing services which they could envisage no possibility of providing themselves. In retrospect, therefore, the 'small-scale' model adopted by the Civils in order to maintain the intimacy and cohesion of its membership was an important contributory factor in the creation of the complicated multiplicity of institutions which has come to characterise the British engineering profession.

Notes

1. Samuel Smiles, *Life of George Stephenson*, 1st edition, 1857, pp. 431-2.
2. Ibid. pp. 439 and 442.
3. ICE *Minutes of Council*, 23 February 1857.

The Beginnings of Institutional Proliferation

4. F. R. Conder, *The Men who built Railways*, Thomas Telford, London, 1983 - a reprint of *Personal Recollections of English Engineers*, London, 1868. The new edition is edited by Jack Simmons. See especially chapter four - 'Demand and supply of engineers' on the shortage of engineers - 'If they were not to be found ready made, they had to be extemporised' - p. 27.

5. ICE *Annual Report* 1846, p. 4.

6. Bernard H. Becker, *Scientific London*, London, 1874; Frank Cass reprint, 1968, p. 111, gives a summary of the evolution of the office of the Secretary.

7. 'The Institution of Civil Engineers' in *The Jubilee of the Railway News*, 1914, pp. 176-179.

8. There was a query about the propriety of this relationship which was raised in ICE *Minutes of Council* on 29 March 1859, but it did not lead to any change.

9. The Institution of Mechanical Engineers, for example, published the first volume of their *Proceedings* in the first year of their existence.

10. ICE *Minutes of Council*, 18 April 1837.

11. Ibid., 19 January 1838. See also: 'The Institution - its origin and progress', in *Journal of the Institution of Civil Engineers*, 1, 1935-36, November 1935, 3-12: 'A class of graduates was established consisting of "persons pursuing a course of study or employment in order to qualify themselves for following the profession of a Civil Engineer"'. No further admissions were made after 1846, when young engineers were admitted as Associates instead - p. 5.

12. ICE *Annual Report* 1838, p. 4.

13. *Journal of the Institution of Civil Engineers, op. cit.* pp. 4-5.

14. ICE *Minutes of Council*, 24 December 1838.

15. Ibid., 2 January 1839.

16. Ibid., 5 October 1841.

17. Ibid., 3 May 1841.

18. For an insight into Gordon's polemical disposition, see the account of his exchanges with the Admiralty about the use of steam power in naval ships: R. A. Buchanan and M. W. Doughty, 'The Choice of Steam Engine Manufacturers by the British Admiralty, 1822-1852', *Mariner's Mirror*, 64, 4, 1978, 327-347.

19. ICE *Minutes of Council*, 26 October 1841.

20. Ibid., 8 November 1841.

21. Ibid., 23-27 December 1844.

22. Ibid., 9 January 1845.

23. See Becker, *op. cit.* p. 112.

24. See note 1 above.

25. ICE *Minutes of Council, 27 January 1863.*

26. Ibid., 21 October 1869.

27. Ibid., 27 October 1869.

28. Ibid., 18 January 1870.

29. Ibid., 31 July 1867.

30. George S. Emmerson, *John Scott Russell - A Great Victorian Engineer and Naval Architect*, John Murray, London, 1977: see especially Chapter 10 - 'the Defence of Honour'. See also R. A. Buchanan, 'The Great Eastern Controversy: A Comment' and the 'Response' from Professor Emmerson in *Technology and Culture*, 24, 1, January 1983, 98-113.

31. ICE *Minutes of Council*, 9 March 1852. The report to which Simmons objected was that in *Proc. ICE*, 9, 1849-50, 181-182, and the paper by W. Fairbairn, 'Tubular Girder Bridges', pp. 233-287.

32. Ibid., 9 March 1852.

33. Ibid., 16 March, 23 March and 6 April 1852.

34. Ibid., 23 June 1846.

35. Ibid., 26 March 1850: Manby's cooperation was requested and permitted.

36. Ibid., 22 May 1855: 'a few young civil engineers competent to aid at the Seat of War in making arrangements for obtaining a better supply of water to the camp before Sebastopol'. Nothing came of this proposal: see 27 June 1855.

37. Ibid., 22 February and 15 March 1871.

38. Ibid., 13 November, 4 December and 11 December 1849.

39. Ibid., 15 November 1866: a letter from E. Noel Eddowes and 92 others.

40. Ibid., 20 November 1866.

41. ICE, *The education and status of civil engineers, in the United Kingdom and in Foreign Countries*, London, 1870

42. See Chapter Nine below for a fuller discussion of this subject.

43. ICE *Minutes of Council*, 20 March 1866.

44. Ibid., For a discussion of the regional societies mentioned here, see below, Chapter Seven.

45. Ibid., 1 July 1867.

46. Ibid., 18 August and 23 August 1880.

47. Ibid., 28 March 1848.

48. Ibid., 10 February and 3 November 1857.

49. Ibid., 1 December 1857.

50. Ibid., 20 September 1858.

51. Ibid., 9 April 1861.

52. Ibid., 5 July 1865.

53. Ibid., 14 November 1867 and 9 March 1869.

54. Ibid., 13 December 1870.

55. The legacy of £2,000 to the Civils is mentioned in ICE *Minutes of Council*, 23 October 1860.

56. I. Brunel, *Life of Isambard Kingdom Brunel*, London, 1870, David and Charles reprint 1971, p. 516, note 2.

57. N. W. Webster, *Joseph Locke - Railway Revolutionary*, Allen and Unwin, London, 1970.

58. For Cubitt, see *DNB* entry.

59. K. H. Vignoles, *Charles Blacker Vignoles- Romantic Engineer*, Cambridge, 1982.

60. E. F. Clark, *George Parker Bidder - The Calculating Boy*, KSL Publications, Bedford, 1983.

61. For Barlow, see *DNB* entry.

62. R. H. Parsons, *A History of the Institution of Mechanical Engineers 1847-1947*, London, p. 10.

63. S. Smiles, *Life of George Stephenson*, 1857, p. 502: the passage is quoted in the anonymous 20pp. typescript, 'A study of an alleged slight' dated 1956 in the ICE Library, which purports to give a full account of Stephenson's relationships with the Civils and to disprove the allegation that he was refused membership.

64. S. Smiles, *Lives of the Engineers*, vol. 3, p. 403, note 2.

65. ICE *Annual Report*, 1849.

66. George Stephenson to J.T.W. Bell at Tapton, Chesterfield, quoted in 'A study of an alleged slight', p. 19.

67. Parsons, *op. cit.*, p. 10.

68. Parsons, *op. cit.*, quoted p. 11.

69. I. K. Brunel, *Private Letter Books*, 6 January 1847: it is not clear if this is a response to the original letter or to a subsequent message.

70. Mechanics Institutes were an attempt to provide useful education to intelligent artisans. They had been pioneered by George Birkbeck in Glasgow and spread throughout Britain in the 1820s and 1830s. Rut they never achieved a high level of education and their appeal was to the upwardly mobile lower-middle classes rather than artisans. See D. Cardwell, *The Organisation of Science in England*, London, 1957.

71. Parsons, *op. cit.*, p. 9.

72. Ibid., p. 20: also IME *Minutes of Council*, 30 April 1856, when the resolution was proposed by Mr J. Fenton of Leeds.

73. ICE *Annual Report*, 1849: as note 65 above: quoted by Professor H. S. Hele-Shaw in his Presidential Address to the IME in 1922 - *Proc. IME*, 1922, p. 1001

74. ICE *Minutes of Council*, 1 December 1857.

75. Parsons, *op. cit.*, p. 17.

Chapter Five

The Fragmentation of the Profession*

The process initiated by the foundation of the Institution of Mechanical Engineers in 1847 became, in the second half of the nineteenth century, one of pronounced institutional proliferation which transformed the British engineering profession from a comparatively small and homogeneous body of men into a large and complex group organised in over a dozen national institutions and a plethora of local societies. The local societies will be the subject of separate consideration in Chapter Seven. The purpose of this and the following chapter is to examine the fragmentation of the profession at the level of the national institutions, to explain why it happened and to examine its consequences.[1]

The apparent homogeneity of the engineering profession before 1847 was by no means perfect. The Institution of Civil Engineers had never included all the practising engineers and had tended to become increasingly selective in its recruitment of new members. No 'missionary' effort had been made by the Civils to assimilate the distinctive new group of railway engineers, even though they provided many of the leaders in the early stages of railway construction. Spokesmen for the Civils had expressed apprehension about the rapid increase in the number of engineers generated by the railway boom, and this attitude did much to encourage the formation of the Institution of Mechanical Engineers as a forum in which the specialised interest of railway engineers could be adequately represented. The result was the establishment of the first institution to disrupt the unity of the profession.

The precedent of an independent specialist organisation became one which was widely adopted in the next sixty years. But the process of institutional proliferation was not one of a simple repetitive pattern. It was complicated by particular personalities and industrial needs, by changing social requirements and by modified perceptions of what could be achieved by the creation of an institution. Analysis of the process reveals a typology of four principal groups of institution.

First, there was the extension of the Civils' model into areas of engineering specialisation which, although essentially new, were adjacent to those for which the Civils already provided professional services. Developments in shipbuilding and marine engineering in the second half of the nineteenth century created areas of this type, and the Institution of Naval Architects was founded in 1860 to cater for the interests of engineers in this rapidly changing and expanding industry.

Second, there was the form of development characterised by the Iron and Steel Institute and the Institution of Mining Engineers, in which old established industries like

---

* See Table I on page 233, *Membership of the British Engineering Institutions, 1850-1914*

the iron industry and coal mining became aware of the need for a more 'scientific' - meaning more systematic, theoretical, and innovative - approach to their trade, and devised institutions to provide this quality. These were industries which had not previously relied heavily on 'engineers' in the usual sense of the term, but the development of the applied sciences of metallurgy, chemical engineering, and mining engineering provided an extension to the scope of engineering which was easily made by managers and others seeking to increase their expertise in these fields, and they readily adopted the forms and procedures of the existing engineering organisations. This category could be further divided between those institutions which, like the Mining Engineers, began as provincial societies and then came together - with substantial safeguards for their autonomy - in federal organisations, and those which, like the Iron and Steel Institute, possessed a more centralised constitution from the start.

Third, some industrial developments in the second half of the nineteenth century were entirely new, and required a different response from the engineering profession. The outstanding example of this type was the burgeoning industry based upon the exploitation of electricity, which began in a very modest way with a few electric telegraph companies but grew steadily as the possibilities of power generation, electric lighting, and electric traction unfolded. Although the expertise for this industry came in the first place very largely from military engineers, academics and scientists, the need for an organised profession of electrical engineers quickly became apparent.

Fourth, there were a few cases - less than might at first have been expected - of splits within existing institutions producing a multiplication of organisations. To some extent, the failure of the Institution of Naval Architects to cater fully for marine engineering promoted the formation of the Institute of Marine Engineers. Similarly, the Institution of Structural Engineers broke away from the Civils because its founders believed that concrete construction required the support of an independent organisation. Usually these 'splits' were fairly amiable, but they were made at the expense of a parent organisation, however much the latter might recoup its losses of membership in other areas.

This typology of engineering institutions is intended to be suggestive rather than definitive. The fact is that no two institutions were exactly alike, but the nature of the circumstances in which they came into existence did much to determine the particular form which they took, so that it is useful to take into account the main variables in these circumstances. The process of institutional proliferation continued from the middle of the nineteenth century to the First World War. It continued, indeed, after the war, although the situation was complicated then by some significant mergers and amalgamations to offset the foundation of new institutions, and there was an increasing emphasis on the need for representative coordinating bodies to speak for the whole profession. There is no obvious pattern in the developments between 1847 and 1914, so in default of any other organising principle it will be convenient to adopt a procedure which is basically chronological, dividing the period into two parts around the year 1880. The earlier period, from 1847 to 1880, was dominated by the formation of institutions in the first two categories, and will be considered in this chapter. The later period, from 1880 to 1914, was distinguished by more new institutions in the third and fourth categories, and will be treated separately in the following chapter.

The third quarter of the nineteenth century was a period of confident growth for the British engineering profession. Carried on the continuing boom in railway construction, the introduction of iron (and later steel) steam ships to replace traditional sailing ships, the increasingly sophisticated engineering requirements of industry, and the vogue for urban improvement with pure water supplies and efficient means of waste disposal, membership of the professional engineering institutions increased almost tenfold from 865 to 8,047 between 1850 and 1880. These were the years in which Robert Stephenson, I. K. Brunel and Joseph Locke - the 'Railway Triumvirate' - reached the culmination of their careers as railway builders. They were also the years of Brunel's last and greatest steam ship, the *Great Eastern*, the largest ship built in the nineteenth century, which was launched after awesome adversities in 1858. And they were the years of J. F. la Trobe Bateman's extensive water works for Manchester at Longdendale and for Glasgow at Loch Katrine, and of Joseph Bazalgette's monumental main drainage scheme created for the Metropolitan Board of Works between 1855 and 1875 to provide London with a desperately needed method of disposing of its organic waste. These were some of the more colourful engineering achievements of the period, but there were many others as engineering expertise was applied increasingly to the search for innovation, higher productivity, and greater safety in industry, and to a wide range of improvements in the standard of life such as gas lighting and better urban transport. Over all, the sense of confident leadership as the 'workshop of the world', evoked so splendidly by the image of the Crystal Palace at the Great Exhibition of 1851, and maintained by two decades of more or less uninterrupted industrial and agricultural prosperity, imbued the British engineering profession with a self-assurance which had not emerged before and which became severely muted later on. In these years from 1850 to 1880, British engineering thus enjoyed a unique euphoria which tended in some respects, such as in its attitudes towards engineering education, to harden into complacency. It is against this background that the fragmentation of the profession has to be seen, as a handful of important groups of engineers followed the Mechanicals in asserting their institutional independence of other engineers.

Although independent, however, the new institutions maintained a striking resemblance to the Institution of Civil Engineers, which provided a model for the innovators. No attempt was made to depart in any significant respect from the form of organisation which had been so successfully pioneered by the Civils. That is to say, the new foundations retained the character of the Civils as organisations of mutual instruction. They were not trade associations and they never entered into negotiations about the salaries and conditions of work of their members, although they occasionally operated some form of benevolent fund. Political and religious matters ware banned from their consideration. They all adopted a hierarchical membership structure modelled on that of the Civils, allowing for a graded ascent from student or junior membership through associate membership to full membership and sometimes to fellowship status. Provision was usually made for grades of honorary or non-practising membership, and institutional business was organised through an elected council of senior members presided over by a President, for whom a one or two year term of office became normal. Most national institutions acquired a small secretariat and club-like premises in central London. All the new creations retained the club-like initiation procedure devised by the Civils, whereby membership was achieved by election amongst existing members on application proposals supported by sponsors and with some sort of demonstration of aptitude. This normally consisted of testimonials from established engineers that the applicant had served a pupilage or performed adequately as an assistant to them. There was no formal examination, and almost until the end of the century no university qualification was considered as equivalent to fulfilling any part of the entry requirement. The emphasis was on practical experience, monitored by professional seniors, and improved by further practice in the light of discussion amongst peers.

Foreign observers have marvelled at the lack of formal requirements for entry into the British engineering profession, but the system of controlled entry into the institutions and the subsequent supervision of professional standards of conduct by the same organisations has served very effectively to provide the controls necessary to maintain the integrity of British engineering as a professional body in the strictest sense of that term. It is true that the institutions have had remarkably little cause to exercise their disciplinary function, but this is a tribute to the self-regulatory character of these organisations, and when called upon discipline has been effectively applied. The model of British professional organisation, devised by the Civils and then replicated many times by new institutions, is thus one of mutual regard for professional excellence maintained by careful control over entry and the supervision of proper standards by a continuing process of collective education.

Just as the Civils acted as a model in the formation of new institutions, so they also provided an evolutionary pattern. Once a new institution had been established, with its own constitution and grades of membership, it aspired to achieve corporate status. Here it encountered a snag, because the Civils remained jealous of their Royal Charter until the end of the century and resisted the attempts of other organisations to receive similar recognition. Many institutions chose, therefore, to seek incorporation under the terms of the current Company Laws, and only subsequently sought a Charter or a Royal Charter when the Civils had abandoned their policy of obstruction. Only tardily did the Civils recognise that they could no longer realistically claim to speak for the whole profession, and that they had become merely the first and oldest amongst equals as far as public pronouncements on behalf of the engineering profession were concerned. None the less, the role of the Civils as a model and an inspiration to all subsequent engineering organisations remained a significant fact throughout the period.

The Institution of Naval Architects was the first national organisation to emerge after the Mechanicals in 1847 and, like the Mechanicals, it fell into the category of institutions which developed to occupy professional niches adjacent to the concerns of the Civils, but too specialised to receive adequate attention in the meetings of that body. It was established in 1860. It celebrated its Jubilee in 1910 by receiving a Charter of Incorporation, and in 1960, for its Centenary, it was granted the privilege of calling itself 'The Royal Institution of Naval Architects'. Of the forty people invited to attend the foundation meeting on 16 January 1860, eighteen managed to do so and resolved unanimously:

'We who are present do now constitute ourselves an Institution of Naval Architects for the purpose of advancing the science and the art of Naval Architecture.'[2]

Amongst the eighteen were Sir Edward Reed, later to become Chief Constructor of the Royal Navy; Dr. Wooley, who had been Principal of the defunct School of Naval Architecture; John Scott Russell, the shipbuilder at whose Millwall yard the steamship *Great Eastern* had been constructed; and John Penn, the eminent London marine engineer. Reed was appointed as the first Secretary, and a formal statement of 'Objects' was approved, which was summarised in the Charter of 1910 as:

'the improvement of ships and all that specially appertains to them, and the arrangement of periodical meetings for the purpose of discussing practical and scientific subjects bearing upon the design and construction of ships and their means of propulsion, and all that relates thereto.'[3]

From the outset it was realised that, although the designation 'naval architect' rightly emphasised the central role of the person to whom the ship as a whole was the proper subject of study, the construction of ships involved an intimate relationship with many specialised skills and crafts. As Reed observed, 'naval architects alone do not produce ships'.[4] So a class of 'Associates' was established which for the first twenty years of the Institution exceeded the number of Members, and from which all the early Presidents of the Institution were recruited - an unusual circumstance for a professional body. The first President was Lord Hampton who, as Sir John Pakington, had been First Lord of the Admiralty and responsible for the construction of H.M.S. *Warrior*, the first British ironclad, in 1859. He held the office for twenty years.

In its choice of Vice-Presidents, also, the Institution differed from other professional bodies by appointing noblemen and gentlemen of eminence who had been closely connected with the Royal or Merchant Navies, with up to thirty (limited to twenty-four in 1890) at any one time serving in a largely honorific capacity.

The list of names published in the first volume of *Transactions* produced by the Institution gave a total of 324, but did not distinguish between Members and Associates. However, out of 365 names classified in 1861, ten were enrolled as Honorary Members and Honorary Associates, sixty-five as Members, and 290 as Associates. Total membership rose to 492 in 1864, but thereafter remained fairly stagnant for a decade, a condition attributed by Sir William White in his Jubilee Review of the Institution to a decline in the excitement which had attended the controversy over ironclads.[5] In 1877 the Institution held a Summer Meeting in Glasgow which demonstrated that the membership was increasing again, the total number of names on the roll rising to 537, with virtually equal representation of Members and Associates. In subsequent years the total membership continued to increase steadily to around 2,100 at the beginning of the First World War, the greater part of this increase being attributable to professional Members with Associates remaining fairly static in numbers, although a new grade of 'Associate Member' was set up in 1899 and proved popular for professional men under the age of thirty (fixed for full Members) and for men over thirty in positions of relatively low responsibility: there were 169 such Associate Members by 1910. Another new grade was created in 1905, when the 'Student' class was opened to suitable young men between eighteen and twenty-five, and there were about a hundred in this class by 1910.

The distinction between 'naval architects' who were Members and other skilled personnel who were recognised as Associates seemed to be generally acceptable to the founders of the Institution and worked reasonably well in practice. It did, however, lead to one point of serious difficulty. This was in relation to marine engineering, the highly skilled business of providing mechanical propulsion for ships. When the Institution was founded in 1860, the rivalry between steam and sail had already been largely determined in favour of the former, even though there was still a considerable mercantile future for sailing ships on the long haul to the Far East, (at least until the introduction of triple compound engines in the 1880s). Nevertheless, at the foundation of the Institution marine engineers 'concurred both in the title selected for the Institution and in the restriction of membership to naval architects, and took their place contentedly in the class of Associates'.[6] But the distinction was an uneasy one, particularly as some founders, such as Scott Russell, practised as both marine engineers and naval architects. So in 1869 the anomaly was removed, with 'marine engineers conversant with naval architecture' being admitted to membership. The corollary of this change - a modification of the title of the Institution to include marine engineering - was also considered, but was rejected. The need for any such change was eliminated in 1889 by the creation of the separate Institute of Marine Engineering.

It does not appear that the foundation of the Institution in 1860 provoked any jealousy from existing organisations, although it is impossible to be absolutely certain on this point as some great shipbuilders like I. K. Brunel were prominent members of the Civils, and the Mechanicals had numbered eminent marine engineers amongst their foundation members. John Scott Russell was himself a Council member of the Civils from 1857 to 1867, and had been a founder-member of the Mechanicals and Vice-President in 1855-6,[7] but he took a leading part in the creation of the new Institution in 1860. Indeed, his organisational expertise was particularly valuable, for in the words of Sir Nathaniel Barnaby, another founder-member, reflecting on the hopes being expressed:

'It needed someone familiar with the working of other great technical Associations to bring such mere wishes to a lively issue. In Mr. Scott Russell a leader was found, and a most capable leader he was.'[8]

Apart from Russell, there was little overlap with the higher echelons of the other two national institutions in 1860, and the new organisation was able to define an area of operation which caused no offense to the others. True, Brunel was dead by then, and it is possible that Russell was motivated in part by an understandable wish to escape from Brunel's shadow in what had been an exceptionally tortured relationship between the two men, but with its commitment to ship construction and its concern for improvements in the education of naval architects, the Institution of Naval Architects gave no cause for anxiety to the senior institutions, which were prepared to give it a cautious welcome.

The really novel feature of the situation was that, with the introduction of iron and later steel as the main materials of ship construction, and with the adoption of steam propulsion and the development of paddle-ships and screw-driven ships, a new area of professional expertise had opened up in the middle decades of the nineteenth century. So large and complex was this field that the existing professional institutions, albeit interested in and sympathetic to the large-scale innovations taking place in it, were unable to give it the full consideration that it deserved and required. The efforts of the experts in this new field to organise themselves for professional purposes was thus thoroughly understandable. It was a process which was to be reproduced several times in the course of the second half of the nineteenth century. Proliferation, in this instance, was a necessary reaction to the expansion of engineering expertise into an area which had been transformed out of all recognition by the adoption of revolutionary new techniques, materials, and processes.

The Institution of Naval Architects thus moved into a professional niche which was not being adequately catered for by the existing institutions, and proceeded to make a highly competent job of providing professional services of organisation, discussion and consultation to persons working in the field. Its *Transactions* began publication in the first year of its existence and established a high standard from the start, reflecting the quality of the expertise represented in its membership and the papers offered for discussion. Volume I contained contributions from Scott Russell on the wave-line theory with which he had established his reputation as a naval architect; from Sir George Airy, the Astronomer Royal, on compass correction in iron ships; from Sir William Fairbairn, a past-President of the Mechanicals, on iron shipbuilding; and from Mr. Lennox on chain cables. Volume II contained the first of the seminal papers presented to the Institution by William Froude on the stability of ships, based on his use of model-testing techniques. As White claimed for the *Transactions*, 'theory and practice both find adequate representation',[9] and there can be little doubt that the regular discussion of such topics greatly enhanced the level of competence amongst British shipbuilders.

Sir William White observed that: 'Before this Institution was founded, naval science had no home in England'.[10] From its beginning, the Institution was anxious to promote this new science. Scott Russell, like many of his contemporary engineers, was a keen advocate of scientific and technical education, and presented a paper to the Institution in March 1863 expressing the views of Council which argued the case for a school of naval architecture and marine engineering. Ironically, two previous attempts to establish such a school, the first in 1811 and the second in 1848, had languished for lack of support; the Principal of the latter establishment, Dr. Woolley, was a founder-member of the Institution and an advocate for a new start in education. A committee was formed and reported promptly, and by the end of 1864 the Royal School of Naval Architecture and Marine Engineering had been organised, with the cooperation of the Admiralty and the Education Department.[11] The School functioned in South Kensington until 1873, when it closed down to be replaced by the Royal Naval College. White regretted that the School never received the support from private British shipbuilders and engineers for which its founders hoped.[12] During its nine years of activity, however, 119 students were under instruction, of whom ninety-five (fifty-three naval architects and forty-two marine engineers) completed the course, and prominent members of the Institution gave their services as lecturers, while there was a staff consisting of young Cambridge graduates, members of the Constructive and Engineering Staff of the Admiralty, and practical shipbuilders and engineers under the Principal, Mr. Merrifield.[13] White was himself a graduate of the

School, so that his treatment of its demise is not entirely objective, but its fate was fairly typical of experiments in scientific and technical education in this period. Only in the last two decades of the nineteenth century did the sense of need become strong enough, and the resources sufficient, to sustain a successful development in this field, The educational break-through came with the development of new university departments rather than in separate educational establishments for specialised subjects. The endowment of a Chair in Naval Architecture at Glasgow University by Mr. John Elder in 1884 was a significant turning-point. The first appointment to this Chair, Professor Elgar, had been at the South Kensington School, and the course which he provided quickly won support. It was soon followed by similar developments at Newcastle and Liverpool.[14]

The Institution became involved, soon after its foundation, in various public activities, advising government departments and other bodies, either through special committees of Council or through nominated representatives. It also developed friendly relationships with foreign societies, several of which were modelled directly upon it, such as the French Association Technique Maritime, founded in 1888; the American Society of Naval Architects and Marine Engineers, incorporated in 1893; the Schiffbautechnische Gesellschaft, founded in Berlin in 1899; and similar bodies in Japan, Sweden, and Italy.[15] Several of these relationships blossomed into international collaboration in Summer Meetings such as the one held in France in 1895, and members of the foreign counterparts were present at the Jubilee Celebration of the Institution in 1911 to add their congratulations to those of other spokesmen.[16] Monsieur L.E. Bertin, President of the Association Technique Maritime, gave a flattering speech and observed that fifty years 'is but a small matter in the existence of Societies such as yours', and he concluded:

'Your fifty-years terms will succeed each other. I take the precaution of congratulating in advance those young colleagues who will celebrate the next one, the centenary of the Institution in 1960, and who will perhaps then say a word or two in memory of their ancestors...'[17]

Three years after the foundation of the Institution of Naval Architects, another group of professional engineers formed an association which was to lead to the establishment of the Institution of Gas Engineers. The first public gas supply had been set up in London in 1812, and by the middle of the nineteenth century it has been estimated that there must have been about a thousand gas works in Britain.[18] Most of these were very small. They were also widely scattered and their technology was fairly primitive, with a handful of experts in the manufacture of coal gas being consulted for advice as necessary. Increasing numbers, organisation and technical sophistication gradually created conditions which were appropriate for institutional development in an area which was peripheral to existing engineering institutions, and which had not previously relied heavily on engineers. This may be regarded as an example of the second type of institutional evolution. The first tentative moves were made in Scotland in the early 1860s, and in 1863 'twenty four gentlemen, being engineers or managers of gas works', met in Manchester and established the British Association of Gas Managers, electing the eminent civil engineer Thomas Hawksley an Honorary Member and installing him as the first President. Another early President, Thomas Newbigging, referred later to the initial meetings in Manchester, when:

'the members, some middle-aged, the majority young men - sat round the table and were animated by one spirit - a spirit of enthusiasm with a modest side to it, but with a determination that the gas industry, of which they presumed themselves to be the first embodied representatives in England, should make some progress through the enlarged intelligence of its members to be brought about by the free interchange of opinion and experience.'[19]

The objective of 'enlarged intelligence' was recognisably that of the professional engineering institutions, but it is not clear how far the forty members of the Association at the end of the first year of its existence were engineers. A proportion certainly consisted of owners and administrators. But the scope of engineering was being steadily expanded at this time, so that it became possible to see the Association as an engineering institution. It held its annual general meeting in a different city each year, and in 1881 the Association changed its name to 'The Gas Institute', creating a new class of Associates and extending its information service to members. By 1888, the Institute had 821 members and six district associations: the examinations in Gas Manufacture introduced by the Society of Arts in 1874 had been taken over by the City and Guilds and had become well established, and the foundations of a Benevolent Fund had been soundly laid.[20]

In 1890, the Institute changed its name once more to 'The Incorporated Gas Institute', but as the total membership had been reduced to 580 it seems that the organisation had run into some difficulties. The nature of these difficulties was indicated by the formation in the following year of 'The Incorporated Institution of Gas Engineers', with about seventy members. The point at issue between the two bodies was that the Institution restricted its membership to professional engineers whereas the Institute accepted businessman and administrators. The estrangement was resolved by an amalgamation in 1902 as 'The Institution of Gas Engineers', but uneasiness remained until 1905 when the formation of 'The Society of British Gas Industries' left the Institution free to concentrate on the professional aspects of gas engineering.[21] Other associations, in Wales and Scotland, subsequently affiliated with the Institution, in 1905 and 1908 respectively. The educational work through the City and Guilds examinations was extended, and a close relationship was established with the University of Leeds through the Livesey Chair in Gas Engineering, and with Imperial College, London.[22]

The technical sophistication of gas production increased significantly at the beginning of the twentieth century, with vertical retorts replacing horizontal and inclined retorts, and with the use of gas for heating and cooking overtaking the function of lighting which had for long been its only use, but which had been seriously challenged by the development of electric lighting. These technical advances confirmed gas engineering as a distinct field of expertise, so that the Institution was able gradually to distance itself from the commercial aspects of gas manufacture which had previously made its identity somewhat elusive. This appears to have been satisfactory as far as its membership was concerned, for by 1914 the total had climbed back to 875.[23] Having been incorporated in 1890, and again in 1902, the Institution was granted a Royal Charter in 1929. It continued to grow in size, reaching 4,000 by the mid-twentieth century.[24]

The third of the engineering institutions dating from the 1860s was the Iron and Steel Institute, founded in 1869.[25] It was not created specifically as a professional association of engineers, but like the Institution of Gas Engineers it came in effect to resemble such a body so closely that it deserves to be included here, particularly as it demonstrated the problem of defining the limits of engineering in a rapidly expanding industry. The British iron industry had been one of the classic 'take-off' sectors of the Industrial Revolution, with the eighteenth century innovations in coke smelting, crucible steel, and the puddling of wrought iron. The industry thereby achieved sufficient momentum to maintain continued expansion throughout the first half of the nineteenth century, following which it underwent a further series of radical technical innovations associated with the development of bulk steel manufacture. Henry Bessemer's 'converter' was first demonstrated in 1856, and the Siemens' regenerative furnace and 'open hearth' process was developed soon after.[26] Then came the 'basic' process of S. G. Thomas and P. C. Gilchrist, first expounded in a paper to the Iron and Steel Institute in 1879.[27] Cumulatively, these processes ushered in the age of bulk production of steel, which rapidly replaced wrought iron for most industrial and constructional purposes. Even more significantly, they thrust the iron and steel industry into a novel situation in which scientific investigation became a necessary part of a successful enterprise, requiring a technical competence which had previously been in short supply.

The formation of the Institute was a response to this need on the part of the iron and steel industry. It was formed on the initiative of a group of Northern ironmasters and many of its early members were the owners and managers of iron works rather than engineers in the normal sense of the term. But it adopted a form which was characteristic of a professional organisation, proceeding through discussion and the exchange of information to test new ideas and to encourage those which promised profitable commercial results, and the proportion of its members who were, in practice if not in name, metallurgists and chemical engineers, grew steadily. The objects of the Institute were summarised by one of its Presidents many years later:

'(the) main objects were to break down technical secrecy and to encourage scientific study and discussion of the operations and possible developments of a great industry.'[28]

Thus, although it began as something resembling an industrial information service, or even an employers' association, the Institute rapidly established itself as a proto-professional organisation, staking out a claim for engineering competence in a crucial modern industry. Its membership grew steadily to reach 2,200 in 1910, and even though it then remained about this level for several decades it underwent a further rapid growth to around 5,000 after the Second World War.

Bessemer had been a member of the Provisional Committee which established the Institute, and became its second President. Siemens had been elected a member in its first year, and became President in 1877. Although neither were, strictly speaking, engineers - Bessemer was more a professional inventor and entrepreneur, and Siemens an industrialist from an intellectually gifted family - their support for the Institute helped to establish its professional credentials as being in the forefront of scientific innovation in the industry, and this position was confirmed in 1879 when Thomas and Gilchrist announced the discovery of their 'basic' process to its members. Early in its life, in 1871, the Institute set up a special committee to investigate the iron ore resources of Britain, and the development of mechanical puddling techniques. Later, in the inter-war years, the promotion of such research activity proved too expensive and led to the formation of the British Iron and Steel Research Association, on the Council of which the Institute retained a strong interest.

Like other professional associations, the Institute gave attention to the promotion of: 'a more systematic cultivation of experimental science',[29] and to the expansion of technical education.[30] Unlike many British professional engineering institutions, the Institute gave a very warm welcome to overseas members, beginning with Thomas Blair, an American ironmaster from Pittsburgh, who became a member in its first year of existence, and including a number of overseas Presidents such as Andrew Carnegie from America, Adolphe Greiner from Belgium, and Eugene Schneider from France. Its highest honour, the Bessemer Medal, was frequently presented to overseas steelmakers and, beginning with a visit to Liège in 1873, the Institute held regular overseas meetings. A substantial proportion of its members - as much as a quarter by 1910 - were from overseas, and although this occasionally caused embarrassment as with the presence of German members during the World Wars, the Institute benefited from this cosmopolitan membership.

The proliferation of national professional institutions for engineers in the second half of the nineteenth century was paralleled by a rash of local or provincial associations, but there seems to have been remarkably little interaction between these and national developments. It is true that the Institution of Mechanical Engineers began with headquarters in Birmingham and went on to become a national organisation, but it had always aspired to be national in its coverage so that there was no change of objectives in this case. For the most part, local associations remained local, with a very narrow focus of interest.[31] The only significant exception from this generalisation was that of the coal mining associations, for although the Institution of Mining Engineers was officially founded in 1889 it was in fact a federation of local institutions, some of which had been established almost forty years earlier.

The first organisation of professional mining engineers in Britain was the North of England Institute of Mining and Mechanical Engineers, established in Newcastle-upon-Tyne in 1852. Its first President, Mr. Nicholas Wood, had anticipated the formation of a wider organisation by expressing the readiness of the Institute to cooperate 'with any institution or society having for its object the prevention of accidents in mines'.[32] The frequent incidence of mining accidents, which cast a grim shadow over the otherwise thriving coal mining industry in the nineteenth century, was a powerful incentive to all people with mining expertise to take some constructive action to control the dangers and the establishment of local professional institutions performed a valuable educational function in this respect. Like the gas and iron industries, coal mining was one of those well established industries which had managed well without professional engineers until changing conditions - in this case the need for greater safety - obliged it to generate the appropriate expertise in the second half of the nineteenth century. The emergence of coal mining engineers as a self-conscious professional group was the direct result of this need.

As local mining institutions were created, the possibility of some sort of federal alliance between them became increasingly realistic. In 1868, Sir George Elliot, Bt. encouraged such plans in his Presidential Address to the North of England Institute,[33] but almost twenty years elapsed before his successor, Mr. John Daglish, revived the notion:

> 'An institute representing the whole of the mining science of Great Britain would be able to supply re-
> liable information to the Government upon the real practical requirements of legislation, and would be
> a power to resist any proposed legislation contrary to the real interest of mine owners and workmen.'[34]

At the time, the Secretary of the North of England Institute was Theo. Wood Bunning, and it was he who took up the campaign by canvassing for an 'Imperial Mining Institute' on a federal basis and with colonial branches.[35] His paper on this subject was referred to a committee, and after further deliberation a meeting was held in Sheffield on 30 January 1889 which proposed a scheme for a federated Institution of Mining Engineers. The proposal was adopted by the North of England Institute and also by the Chesterfield and Midlands Counties Institution of Engineers; the Midland Institute of Mining, Civil, and Mechanical Engineers; and the South Staffordshire and East Worcestershire Institute of Mining Engineers. The Institution of Mining Engineers 'was accordingly founded by mutual agreement of these institutes' on 1 July 1889.[36] Other local bodies joined later: the North Staffordshire Institute of Mining and Mechanical Engineers in 1891, and the Mining Institute of Scotland in 1893. Despite the grandiose titles of some of these local bodies, their membership consisted overwhelmingly of coal mining engineers.

The first home of the Institution was Neville Hall, headquarters of the North of England Institute in Newcastle, provided by the local association 'free of all charges', but in 1907 the offices were moved to London. The *Transactions* first appeared in 1889, edited by M. Walton Brown, the Secretary. They consisted largely of papers submitted to the constituent bodies of the federation. John Marley of Darlington was the first President, and the Institution had an initial membership of about 1,150 derived from its member societies.[37] This figure grew steadily to 2,486 in 1899 and to 3,161 in 1909.[38] By 1914 it had reached 3,277.[39] The constituent bodies, however, remained largely autonomous in possessing their own rules (including even their own charters in some cases), collecting their own subscriptions, and holding their own meetings. The powers of the federal organisation were thus seriously curtailed. But at least one President was prepared to make extensive claims for it; in his Presidential Address of 1938-9, Sir Frank Simpson said:

> 'I think we are justified in claiming that through the instrumentality and public-spirited labours of the
> Institution and its numerous committees, and those of the Federated Institutes, enormous improvements
> have been effected in the technical operation of coal-mines, and in safeguarding the health and general
> well-being of mine-workers.'[40]

The Institution was incorporated by Royal Charter in 1918. In 1969 it was de-federated, the previously autonomous constituent bodies becoming branches in what had thus become a centralised national organisation.

The last national organisation to be considered from the period 1850-1880 arose out of the needs of urban engineering. The need for improvements in public services such as highways, water supply, and waste disposal, had been widely recognised in Britain in the first half of the nineteenth century, when towns and cities had begun to expand at a phenomenal rate in response to the stimulus of rapid industrialisation. There had been little agreement, however, about the best way of providing these services for the urban communities, except that the conventional wisdom of the period accepted the premise of classical political economy that such things should be left to individual enterprise in pursuit of private profit. The Towns Improvement Act of 1847 and the Public Health Act of 1848 marked a significant change of view towards a greater reliance on the municipal corporations in providing the services which had become essential in order to curtail the epidemics of such devastating diseases as cholera. These measures encouraged the appointment of a 'Town Surveyor', and the Public Health Act also created a General Board of Health which, amongst other things, was given the authority to prevent the dismissal of a Town Surveyor without its prior sanction. By the second half of the century, therefore, the more enterprising towns and cities had begun to appoint appropriate officers, including surveyors and sanitary engineers - the nomenclature was sometimes vague and variable - and had begun to improve the services on which the health and well-being of the population depended. This new commitment to urban renewal was not seriously interrupted by the abolition, in 1858, of the General Board of Health, for reasons which were political rather than environmental. It did, however, have the consequence that the newly emerging group of municipal engineers were deprived of the measure of protection from the interference of their elected masters which the 1848 Act had provided for them, and there is no doubt that the surreptitious pressure placed on them by contractors and other interested parties was increased as a result.[41]

This had become a rather serious anomaly by 1871, when the Royal Sanitary Commission reported. It stated clearly the problem of maintaining the momentum of urban improvement:

> 'Great is the *vis inertiae* to be overcome, the repugnance to self-taxation, the practical distrust of science, and the numbers of persons interested in offending against sanitary laws, even amongst those who must constitute chiefly the local authorities to enforce them.'[42]

But the commissioners did not bother to consult the municipal engineers, and one of their major recommendations, which was rapidly translated into law, was that a new class of officer - a Medical Officer of Health and an Inspector of Nuisances - should be appointed with the same sort of protection against arbitrary removal which had previously been enjoyed by the Town Surveyors. The latter, mostly now regarding themselves as municipal engineers, were understandably distressed and were stirred to protest at this virtual down-grading of their status. The Institution of Municipal Engineers emerged out of this protest.

The lead was taken by Mr. Lewis Angell, Engineer to West Ham Local Board. On the appointment of the Royal Sanitary Commission, Angell had written to the Chairman, Sir C. B. Adderley, observing that local surveyors and engineers needed to be protected if they were to carry out their duties according to principle rather than policy:

> 'The Surveyor of a Local Board ought to be protected in a similar manner to the Poor Law officers, and there should always be an appeal to some power above, say the Local Government Act Office.'[43]

A largely one-sided correspondence ensued, in which Angell repeated his point firmly and clearly, receiving mainly evasive answers from the official spokesmen. It became apparent that the Government had no intention of re-introducing the sort of legislative protection which the municipal engineers desired, so on 29 April 1871 Angell convened a Conference of Borough Engineers and Town and Local Board Surveyors at the offices of the Institution of Civil Engineers, at which the desirability of a permanent association was considered and a committee appointed to continue the campaign. This committee made a deputation to the President of the newly-formed Local Government Board, Mr. James Stansfield, MP, on 8 February 1872. Stansfield was not unsympathetic, but he made it clear that he thought the objectives of the engineers would be best accomplished by the 'improved constitution of local authorities' and by the spread of education. The deputation was not convinced and determined to establish its own organisation. But twelve years later, Angell was prepared to admit that:

> '... some of Mr. Stansfield's hopes have been realised. Wonderful progress has been made, and the position of local officers both professionally and financially immensely improved, so that some of us holding larger appointments are not so anxious for protection as we were, thanks to an improved appreciation of our services.'[44]

The 'improved appreciation' to which Angell referred was due, in part at least, to the success of the organisation which he established in presenting a united and impressive public image for the profession of municipal engineers, and Angell himself deserves much of the credit for this achievement. The 'Association of Municipal and Sanitary Engineers and Surveyors' was established at a meeting at the Civils in February 1873. The inaugural meeting was held on 2 May, and Angell was elected the first President. He held the office for two years, after which it became an annual appointment. The objects were given as:

> '(a) The promotion and interchange among its members of that species of knowledge and practice which falls within the department of an Engineer or Surveyor engaged in the discharge of the duties imposed by the Public Health, Local Government and other Sanitary Acts.
> (b) The promotion of the professional interests of the members.
> (c) The general promotion of the objects of Sanitary Science.'[45]

Membership was limited at first to 'Civil Engineers and Surveyors holding permanent appointments under the various urban and rural sanitary authorities within the control of the Local Government Board', assistant surveyors being excluded to begin with. The Association was incorporated in 1890. It amalgamated with the Scottish Association of Municipal Engineers in 1904, affiliated with the County Surveyors Society of Ireland in 1913, and underwent several other modifications of constitution and title until 1948, when it received a Royal Charter under the title of 'The Institution of Municipal Engineers'.

Two points deserving special comment about the Institution concern its membership and its examinations. The circumstances of its creation ensured it a solid 'constituency' amongst engineers serving local authorities, and it began with a membership list of 180.[46] It did not increase much in its first decade, but by the time of its incorporation in 1890 it had reached 403 and it continued to climb steadily to 1,583 in 1912-13.[47] The increase was assisted by the creation of a Graduate class in 1886, a class of Associates in 1902, and a class of Students in 1909. A significant feature of its membership was the high degree of over-lap with membership of other professional institutions. A recent estimate was that one third to one half of the membership held dual membership with the Civils, and a considerable number were also members of the Structurals.[48] It seems likely that such dual membership, or even multiple membership, has been a feature of the Institution since its outset, with Angell and other leading founders having been members of the Civils.

As far as examinations are concerned, the Institution of Municipal Engineers claims to be the first engineering institution to have established its own examination system in the shape of the 'Testamur' examination in 1886 (literally, a certificate that one has passed an examination). This was introduced because it became important in selecting candidates from a wide range of preliminary qualifications, or none at all, as suitable representatives of the profession in the specialised conditions or local government service. At first, the qualifications of membership of the Civils and certain other prerequisites were accepted, but gradually it became desirable to achieve uniformity of practice by insisting on satisfactory performance in the Institution's own examinations. Thus the 'Testamur' came to be increasingly awarded internally by success in these exams. A commentator in the 1950s observed that the Testamur:

'has developed into its present form: a professional examination by professional men for the profession of municipal engineering. It is now accepted throughout the country and Commonwealth as a standard qualification.'[49]

The Institution thus performed an important pioneering role in the educational field, and helped to make some form of examination acceptable as a prelude to membership of an engineering professional association. The gradual acceptance of this principle assisted the transformation of a putative 'gentlemen's club' into a businesslike professional body equipped to look after the interests of its members.

The history of the Institution of Municipal Engineers is thus a substantial success story. Largely by its existence, firmness, and discrimination, it was responsible for converting the office of the local government engineer from that of a general dogsbody of the elected councillors into a highly responsible professional officer whose integrity was generally above suspicion. Angell, in his Presidential Address of 1873, had observed:

'our profession is almost daily insulted by advertisements requiring the knowledge of an Engineer and the administrative ability of a Manager, for the pay of a mechanic.'[50]

And he had been scathing about the not untypical contrast between advertisements for Medical Officers at £800 p.a, alongside those for Surveyors at £150 p.a.:

'Such advertisements lead the public to infer an inferior social status and qualification of a profession which can produce an unlimited supply of pseudo-Surveyors.'[51]

But twelve years later he was able to congratulate the profession on having escaped subservience to 'a transient board of jerry builders' - hardly a flattering description of the local elected council - and despite occasional storm clouds, as in 1905, when it was found necessary for the Institution to express public disapproval of a construction company which had offered a commission on orders obtained to members of the Association,[52] the Institution rarely needed to assert its independence and professional integrity. It should be classified with those institutions like the Mechanicals and the Naval Architects which, developed to occupy a professional niche as a branch of civil engineering, acquired a specialised quality beyond that which fell within the normal compass of the Civils. Like other institutions in the engineering profession, it placed a high value on the educational function of its activities, and it pioneered the introduction of an examination system for its members. Rather more than some other institutions, it was acutely aware of the need to protect the status of the municipal engineer, and in this respect it became a discreet but highly successful professional pressure group.

The period from 1850 to 1880 was thus one of considerable growth for the British engineering profession, accompanied by a distinct spirit of self-assurance such as it had not enjoyed before and has rarely enjoyed since, while industrialisation proceeded in an economy which was generally buoyant. But the professional euphoria engendered by this success obscured the significance of a development which was beginning to impair the coherence of the engineering profession, because the creation of a series of flourishing institutions to deal with new fields of specialisation like naval architecture, or to bring new scientific skills to traditional fields like coal mine engineering, progressively weakened the ability of the Civils to represent the whole profession and failed to put any other unifying authority in their place. The result was a gradual fragmentation of the profession, the deleterious effects of which were not immediately apparent in a period of economic expansion, but which began to take a toll on professional self-confidence in the more difficult decades after 1880. By this time, however, the process of institutional proliferation had gathered momentum, and it was all set to continue the fragmentation of the profession.

Notes

1. A shorter version of the argument in these two chapters was published as 'Institutional Proliferation in the British Engineering Profession, 1847-1914' in *Econ. Hist. Review*, 2nd series, 38, 1, February 1985, 42-60.

2. Sir William White, *The History of the Institution of Naval Architects and of Scientific Education in Naval Architecture*, 1911, p.5.

3. Charter of the Institution of Naval Architects, 1910.

4. Quoted White, *op. cit.*, p.7.

5. Ibid., p.16.

6. Ibid., p.14.

7. George S. Emmerson, *John Scott Russell*, 1977, p.178.

8, Quoted White, *op. cit.*, p.12.

9. Ibid., p.24.

10. Ibid., p.28.

11. Ibid., 19-20.

12. Ibid., p.21.

13. Ibid., p.20.

14. These developments are discussed below in Chapter Nine.

15. White, *op. cit.*, p.18.

16. Ibid., 28-33, which gives an account of the discussion of White's paper.

17. Ibid., p.30.

18. A. E. Haffner, 'Centenary Presidential Address' in *Jnl. Inst. Gas Engs.* 3, 1963, 355.

19. Quoted Haffner, *op. cit.*, p.355.

20. Ibid., p.357.

21. Ibid., p.357.

22. Ibid., p.357. See also *Trans. Inst. Gas Engs.*, 85, 1935-6, 790, where cooperation with the universities is discussed in the presidential address of Colonel W. Moncrieff Carr.

23. *Trans. Inst. Gas Engs.*, 11, 1914, p.73.

24. A substantial account of the Institution has been written to celebrate its centenary by W. T. K. Braunholtz, *The First Hundred Years 1863-1963*, Institution of Gas Engineers, London, 1963.

25. Note the decision to use the term 'Institute' rather than 'Institution' in the title. Most of the nineteenth century professional organisations of British engineers chose to use the word 'Institution'. I have not managed to discover any reason for this choice in preference to 'Institute' as adopted by a minority of them. No particular statement appears to have been intended by the choice of one rather than the other. But the two words are not exactly synonymous in contemporary usage, although the OED suggests no useful distinction between them (Fowler's *Modern English Usage*, on the other hand, does have an entertaining paragraph illustrating different nuances of the two words). 'Institution' has acquired more of the implication of an independent, autonomous, body, whereas 'Institute' carries one or other of several more particular connotations: as an educational body, such as a 'Mechanics' Institute'; as a specialised unit within a larger body, such as an 'Institute of Archaeology' within a university; or as a body with some specific policy function such as an 'Institute of International Affairs'.

26. For a standard account see David S. Landes, *The Unbound Prometheus*, Cambridge, 1969. See also Henry Bessemer, *Sir Henry Bessemer FRS - An Autobiography*, London, 1905.

27. *Jnl. Iron and Steel Inst.*, 1, 1879, 120-134.

28. Richard Mather in his 1951 Presidential Address, *Jnl. Iron and Steel Inst.* 168, 1951, 121-5: the quotation is from p.121.

29. Mather, *op. cit.* p.123, quoting the first President.

30. J. Jones, *Jnl. Iron and Steel Inst.*, 6, 1876, 342-6.

31. These local associations are discussed below in Chapter 7.

32. Quoted in 'History of the Institution of Mining Engineers', in *Trans. Inst. Mining Engs.*, 12, 1898, p.6.

33. Ibid.

34. Ibid. This is an indirect quotation of Daglish's words.

35. Ibid., quoting from a paper read on 6 August 1887: see also *Trans. Inst. Mining Engs.*, 1, 1889-90, xiii-xxxv - 'Notice of preliminary proceedings... '

36. Ibid., p.7.

37. *Trans. Inst. Mining Engs.*, 1, 1889-90, xiii-xxxv.

38. *Trans. Inst. Mining Engs.*, 38, 1909-10, 484, presents this information in tabular form.

39. *Trans. Inst. Mining Engs.*, 48, 1914-15, 469.

40. *Trans. Inst. Mining Engs.*, 97, 1938-39, 13.

41. See 'Evidence of Surveyors... ' in the 'Appendix: Administration of the Sanitary Laws', assembled by Lewis Angell and published in *Proc. Assoc. of Municipal and Sanitary Engs. and Surveyors*, 1, 1873-74, 244-8.

42. Ibid., p.221.

43. Ibid., p.227, letter of 18 February 1870.

44. Lewis Angell, 'Origins Constitution and Objects', in *Proc. Inst. Mun. Engs.*, 11, 1884, 8-17.

45. Summarised in a note on 'Foundation of the Institution', in Inst. Mun. Engs. *List of Members*, 1972, p.2.

46. 'List of Members' in *Proc. Inst. Mun. Engs.*, 1, 1871. By restricting membership to engineers and surveyors for local government boards, those working for the Metropolitan Water Board appear to have been excluded. Sir Joseph Bazalgette, for instance, only became an Honorary Member.

47. The figures are available in the annual volumes of *Proc. Inst. Mun. Engs.*

48. Verbal communication with the Assistant Secretary, 2 May 1980.

49. Percy Parr, 1954 Presidential Address, in *Proc. Inst. Mun. Engs.*, 81, 1954-55, 5. Parr went on to observe, apropos the proliferation of technical qualifications, not all very reputable: 'The time would appear to be more than ripe for some action to limit the number and activities of unnecessary associations and institutions' (p.8).

50. *Proc. Inst. Mun. Engs.*, 1, 1873-74, 24.

51. Ibid., p.21.

52. Parr, *op. cit.*, p.5.

Chapter Six

Divided Voices - Fragmentation Continued

British engineering continued to flourish in the period from 1880 to 1914 as it had done in the previous three decades, but the profession began to show signs of serious disturbance which caused a diminution in its exuberant self-confidence. Symbolically, the collapse of the Tay Bridge in December 1879 marked an important turning point, because it both dealt a body-blow at the self-esteem of a profession which had begun to feel that it could do no wrong, and served to support the campaign of those who sought radical changes in technical education. For a generation, the engineering profession had confidently maintained that its method of training by practice and pupilage was adequate, as shown in the complacent *Report* issued on the subject by the Civils in 1870.[1] The designer of the Tay Bridge, Sir Thomas Bouch, knighted on its opening in 1877, had been a prominent advocate of the traditional engineering virtues of practical experience and economy of resources. The disaster on the night of 28 December 1879, when a complete train with all its passengers disappeared as a result of a central section of the bridge collapsing into the sea during a high gale, discredited Bouch and caused reflection on the theoretical inadequacy of the structure. The subsequent Enquiry concluded that the bridge was 'badly designed, badly constructed and badly maintained', and that Bouch was mainly responsible for these defects.[2] Bouch protested that he had taken the best possible advice, but his career was ruined and the old tradition of empirical engineering stood condemned with him. In future, engineers would he expected to understand the detailed theoretical aspects of their structures. The fact that the great multi-cantilever steel bridge built by Fowler and Baker to carry the railway over the Firth of Forth in 1890 incorporated a massive safety factor in its design was an understandable precaution, and the success of the Forth Bridge went some way towards restoring the tarnished image of the profession. Yet conditions could never be quite the same again after 1880, and the profession became more attentive both to its own educational spokesmen in the universities and elsewhere, and to the campaign for more and better technical education. The result was a substantial increase in the availability of technical instruction, and an increasing readiness on the part of the engineering institutions to recognise technical and university qualifications as appropriate steps towards admission and progress through their grades of membership. In 1897 the Civils, after long deliberation, adopted regulations specifying the recognition of such procedures, and the other institutions gradually followed.[3]

In many ways, the three decades before the First World War continued the engineering developments of the previous period. Railways still absorbed the attention of a large number of British engineers, both at home and abroad, as did shipbuilding, the iron and steel industry, and coal mining. Similarly, the vogue for urban improvement which had begun in mid-century rose to a crescendo as Victorian cities equipped themselves with massive water works, public buildings, electric tramways, and other services, and all of these required the attention of engineers. But there were also innovations and cross currents which contributed to the dispersal of the previous euphoria. For one thing, the universal supremacy amongst prime movers enjoyed by the steam engine since the 1840s began to be challenged, both by internal-combustion engines and by the application of electricity. In neither of these new technologies did British engineering enjoy the unchallenged leadership which it had monopolised in steam technology, so that it had to adopt innovations made in Europe and America. In the manufacture of automobiles and aeroplanes, as in chemical engineering and, outstandingly, in the generation and application of electricity, British engineering, like the British economy generally, came under increasing pressure from overseas competition. Economic historians have spoken of the 'Great Depression' to describe the period of disruption in British industry at the end of the nineteenth century.[4] There was, however, no equivalent disruption in the growth of the engineering profession, the membership of 8,047 in eight national institutions in 1880 being multiplied five-fold to 40,375 in seventeen institutions by 1914. Nevertheless, there were signs that the profession was reacting to the times with increasing uncertainty. New institutions continued to be formed, although several of them were now of the type which arose in response to the demands of new industries such as electricity generation, and, much more than previously, by conscious break-away movements from existing organisations. But the proliferation of institutions was no longer accepted as being as desirable and inevitable as it had been previously. Anxiety began to be expressed about the loss of cohesion in the profession, with a diversity of voices instead of a single united voice, and a concern emerged to reverse the process of fragmentation. Engineering generally, although still flourishing and successful, began to seem a divided and increasingly anonymous profession. This was the context within which the continuing institutional proliferation of the profession occurred between 1880 and 1914.

The most important institutional development of this period was the emergence of the Institution of Electrical Engineers. The Electricals, one of the largest of present-day engineering institutions, derived from the Society of Telegraph Engineers, which had actually been founded in 1871, although it is more appropriate in this context to regard it as a post-1880 foundation. On 17 May 1871, eight men had met at 2 Westminster Chambers, London:

'to consider the expediency of forming a Society of Telegraph Engineers, having for its object the general advancement of Electrical and Telegraphic Science, and more particularly for facilitating the exchange of information and ideas among its Members.'[5]

The official historian of the Institution has observed that there is conclusive evidence that the plans to take such action had been widely discussed for a considerable time, and this was demonstrated by the way in which the eight founder-members present at the initial meeting proceeded 'with complete agreement and knowledge of their objective... to dispatch the whole business of inauguration.'[6]

In fact, the foundation was long overdue, and had been anticipated as early as 1837 by the establishment of an Electrical Society in London for 'the experimental investigation of electrical science in all its branches, and its advancement, not only by pursuing original paths but by testing the experiments of other enquirers'.[7] The moving spirit who presided over this Society had been William Sturgeon, a remarkable autodidact who had begun his working life as a cobbler but had become a passionate experimenter and innovator of electrical devices. The Society had been supported by Faraday and other leading practitioners in what was still a new field of scientific enquiry, and it had undertaken a publishing programme for papers and abstracts which had proved to be over-ambitious, because it ran up a debt which, although modest, persuaded the members in 1843 that it should be terminated. Its Hon. Secretary at that time was C. V. Walker, who accounted for the demise of the Society as having 'succumbed to chronic atrophy'.[8] Walker himself went on to become President of the Society of Telegraph Engineers in 1876, but otherwise there does not appear to have been any significant continuity of personnel between the two organisations.

Despite the collapse of this promising initiative, the middle decades of the nineteenth century were a period of exciting progress in the application of electricity. The possibility of harnessing an electric current had arisen when Volta made his first 'voltaic pile' in 1800, but the exploitation of this possibility took some time. Faraday's demonstration of the additional possibility of the mechanical generation of its power and its utilisation to produce motion came in 1831, before much progress had been made. And then the first significant breakthrough in the commercial use of electricity came with Cooke and Wheatstone's successful experiments in electrical telegraphy. The new technique was promptly installed as a signalling device by I. K. Brunel on the Great Western Railway where it achieved fame by its success in leading to the arrest of a wanted criminal in 1845.[9] Thereafter it was widely adopted on the expanding railway networks, and was quickly applied to commercial and political communication in the form of cable telegraphy. The electric cable linked England and France in 1851, and schemes were immediately promoted to lay trans-oceanic cables. After several disappointments, a cable was successfully laid across the Atlantic under the direction of Daniel Gooch on board the S.S. *Great Eastern* in 1866, and similar projects to the Far East and Australia were then undertaken with enthusiasm. In 1868, the British Government took steps to acquire a monopoly control over the telegraph services, and this took effect in 1870. The number of telegraph offices open to the public in Britain rose from about 2,000 in that year to 5,600 five years later, and in the same period the number of messages transmitted in Britain increased from six to twenty millions.[10] Electric telegraphy had clearly arrived as big business, demanding a high level of professional competence on the part of the large number of engineers who were required to establish and maintain the service.

The need for some form of professional organisation in this new field of engineering skill was manifest, but the existing institutions showed little interest in it and with the intense involvement of the Government it is not surprising that the initiative for the formation of the Society of Telegraph Engineers in 1871 came largely from military engineers. Of the eight founders of the Society, five were service men: three were officers of the Royal Engineers (Major Stotherd, Captain Webber, and Captain Malcolm), one was a Naval officer (Captain Colomb), and Major Bolton, who acted as the first Secretary and eventually became Sir Francis Bolton (1831-1887), had served in the Royal Artillery and infantry. The other three members of the founding fraternity were Wildman Whitehouse, Louis Loeffler, and Robert Sabine. Whitehouse, who took the chair at the inaugural meeting, had already achieved distinction as a submarine telegraph engineer and had been associated with a report on the subject to the British Association in 1855, and with a discussion on it at the Institution of Civil Engineers in 1857.[11] Sabine was a consulting engineer with experience of cable laying: he also managed a factory for his father-in-law, the veteran electrical engineer Sir Charles Wheatstone, and was the author of important books on telegraphy. Loeffler acquired a considerable personal fortune from his association with Messrs. Siemens Brothers and Co., one of the first firms in Britain to recognise the huge scope offered for the manufacture of electrical equipment.

These eight were included in the 'List of Original Members' entered in the Minutes of Council for 17 May 1871, containing a total of sixty-six names amongst which were two Siemens, two Preeces, three Varleys, several Royal Engineers, Lord Lindsay, Frank Scudamore, Bruce Warren, and Latimer Clark. At the second meeting, on 31 May, a ballot was held: Charles Siemens was elected President, and Lord Lindsay and Frank Scudamore became Vice-Presidents; Bolton was confirmed as Hon. Secretary and Sabine was elected 'Treasurer and Librarian'. A Council of eleven contained one FRS (Professor G.L. Foster), three Royal Engineers, and three Civil Engineers: in addition, Siemens was both FRS and CE. The Institution of Civil Engineers was requested to provide rooms for meetings, presumably on the strength of the dual membership of Siemens and other Council members, and after some delay the Civils' Secretary James Forrest replied on 7 February 1872 granting this request and adding a warm welcome from Council to the new Society, hoping 'that it may grow to be a very successful and prosperous body'.[12] This facility was provided 'free of all expense', and as the Annual Report of the Electricals for 1873 observed 'this...was as great a service as could possibly be wished for by a new Society'.[13]

The welcome of the Civils, which was similar to that given to several other new engineering institutions, was reflected in the respectful way in which the Electricals chose the older body as the model for their own constitution. The Electricals were an institution of the type developing in a completely new field of engineering, but they showed as much tenacity as older types in adhering to the well-tried pattern of the Civils. Bolton had circulated some proposals for rules on this pattern to interested parties as early as 9 August 1870, and this helps to explain why the new association came into existence so smoothly, because the proposed organisation was already familiar and acceptable to many members. Latimer Clark, CE, President in 1875, seems to have played a large part in fashioning the new organisation in the image of the Civils. The Society was thus established with Members, Associates, and Honorary Members, with Students and Foreign Members soon being added. By the first Ordinary General Meeting in February 1872, total membership had already reached 110, not including Foreign Members. A year later it had risen to 352, which included twenty-five Foreign Members, and growth continued rapidly to 2,100 in 1890 and 7,000 in 1910. Support from the military services was well sustained, and the Society also received a significant recruitment from the scientific and academic communities, including Fleeming Jenkin, William Thomson, Charles Wheatstone, Clerk Maxwell, G.B. Airy, and Lyon Playfair. As the official historian of the Society observed: 'with such support, the Society could not fail'.[14]

Nevertheless, the Society did have some financial anxieties in its early years, associated with the publication of its *Journal* - the same issue as that which had caused the failure of the Electrical Society in the 1840s.[15] Helped, however, by a donation from the first President and some sound book-keeping, these difficulties were being overcome by 1877. A full-time Secretary had relieved Major Bolton of most of the routine work of running the Society by 1872, when first Schütz Wilson and then George E. Preece held the post. The Society considered applying for a Royal Charter, but decided against doing so in 1880 when the opposition of the Civils to such a move was recognised. Instead, it sought registration under the Companies Acts and was incorporated in 1883.

When the Society was formed, electric telegraphy was the outstanding and highly successful application of electricity, so it was understandable that this was the dominant preoccupation of the organisation in its early years. After 1880, however, new developments began to emerge and to change the balance of interests amongst the membership. In particular, the introduction of the mechanical generation of electricity on a commercial scale to provide power for lighting and transport systems created a major new field of engineering employment, and it was natural that the Society should want to include electrical engineers working in this area and other areas besides that of electric telegraphy. Concern soon began to be expressed, therefore, as to the adequacy of the title of the Society. In 1879, Council considered a letter from Latimer Clark regretting the exclusion of 'that large and increasing body of Electricians who have no connection with telegraphic engineering or the practical applications of Electricity, but who pursue the subject from a pure love of science...'.[16] Council agreed, and discussed variations of the title to include the word 'Electricians'. Eventually, however, W. H. Preece came up with the most satisfactory solution. In 1887 he proposed the adoption of the title 'Institution of Electrical Engineers',[17] and at the same time R. E. Crompton warned members that the unrepresentative character of the Society might 'lead to the formation of a new and rival society'. After further deliberation, Preece's suggestion was accepted in 1888, and the Institution thus assumed the name by which it has subsequently been known.[18]

The source of Clark's anxiety in 1879 had been less with the heavy power engineers, who only became a problem in the 1880s, than with exponents of 'a pure love of science' whom he feared were being excluded from the Society. Similar concern had been expressed in 1873 when the 'Physical Society' had been formed to promote scientific experiments into electrical phenomena, with the support of some of the leading physicists and electrical scientists of the day. It had been felt that this organisation would be a rival to the Society in the search for new members, and Major Bolton had argued in the Council that an amalgamation should be sought with the new body. But those taking a longer view of the division of functions between the two organisations came to see it as an advantage to both to have a 'courteous wall' of neighbourly restraint between the natural scientists and the professional electrical engineers.[19] Certainly the separation confirmed the function of the Society as a professional institution, diminishing its aspiration to be a

scientific society while preparing it for its subsequent development in power engineering and other new aspects of electrical engineering, and without having any adverse effect upon its membership.

In 1878, F. H. Webb was elected Secretary of the Society, and Professor W. E. Ayrton,
recently returned from a post as a teacher of electrical engineering to students in Japan,
took on the chairmanship of the Editorial Committee responsible for the *Journal* and
other publications. Ayrton was one of the 'professorial contingent' which had quietly asserted a strong academic influence on the contents and quality of the papers of the Society.[20]

One claim to distinction on the part of the Society was that it was the first engineering institution to consider admitting women to its meetings. Major Bolton had proposed
admitting Mrs. Lundy as an Associate in 1873,[21] but Council decided that it would be
prudent first to consult the 'parent institution' as they met on the Civils' premises. The
Civils responded that it would not be appropriate to admit ladies to their lecture theatre,
so the matter was dropped for several years, as the Electricals did not acquire their own
premises for meetings (they had their own offices from the outset) until 1910. But the
fact that telegraphy involved a large number of lady operators, many of them from homes
of impeccable social respectability, meant that the pressure for their admission would be
renewed. Captain E. D. Malcolm RE, one of the eight founders, claimed credit for re-
pelling the female invaders:

> 'There was a great desire to increase the funds (of the Society), and it was proposed to turn this necessity
> into an opportunity for admitting female telegraph operators as members. Then it would have become
> a mere dot-and-dash Society. I strenuously opposed the proposal and saved the Society, which is now
> (1918) one of the most important technical institutions in the world.'[22]

This claim specifies the apprehension, probably general amongst Victorian professional
men, that the admission of women would have reduced the status of the Institution from
that of a high-class gentlemen's professional club to that of an amateur enthusiast's organisation.

While the Institution of Electrical Engineers went on to become, in the course of a
few decades, one of the most successful of British professional engineering associations,
the majority of the institutions which emerged in the period 1880-1914 remained fairly
small. Several of them had the distinctive qualities of break-away type associations, while
others developed to occupy small but highly specialised professional niches. Whatever
the origin, however, the Civils continued to provide the model for institutional organisation.

The Institute of Marine Engineers began in East London, with a meeting in the Work-
men's Hall, West Ham Lane, Stratford, on 12 February 1889. This meeting was convened
by thirty-three men who had formed an 'Organising Committee', partly in reaction to the
establishment of the Marine Engineers' Union in 1887. This later organisation was a trade
union, 'having as its objective the recognition of certificated sea-going engineers as being
entitled to a status equal to that of their brother deck-officers, with corresponding improvement in their conditions of service and emoluments'.[23] The members of the Organising Committee did not disagree with the desirability of achieving equality of status for

engineers amongst the crews of ships, which had become a topic of increasing concern and resentment with the rapid transition from sailing ships to steam ships, a process which had been accelerated in the 1880s by the successful adoption of 'triples' in place of simple compound engines. But they sought to achieve the objective by raising the intellectual and social standards of marine engineers rather than by bargaining with other parties. In the circular letter which they distributed to the chief engineers of all steamers in the Port of London on 20 November 1888 they stated that:

'It has been considered highly desirable to form an Institute and Club for Marine Engineers, having for its objects the promotion of those intellectual and social qualities which render life more honourable and pleasant by their possession.'[24]

The invitation was well prepared by the Organising Committee under the Chairmanship of Mr. Asplan Bedlam, who later became the first President, with Mr. James Adamson as Honorary Secretary, and it received a warm response. By 1891, the second volume of the *Transactions* of the Institute already recorded a membership of 452. By 1901 it had risen to 938 and by 1915 to 1,467.[25] It was clear that the sea-going fraternity of marine engineers - the men who actually attended to the engines of ships, as distinct from the manufacturers of marine engines who were already eligible for membership of the Institution of Naval Architects as well as the Institution of Mechanical Engineers - represented a *lacuna* in the coverage of professional organisations, and that the new association thus fulfilled a genuine need. It seems likely that this group accepted a certain social distance from the existing professional institutions, possibly reflected in their choice of the designation 'Institute', with its overtones of mechanics' education and social clubs, rather than the more dignified 'Institution'.[26] Unlike some of the professional engineering consultants, the marine engineers remained close to the engines for which they were responsible. A high proportion of their members were always at sea, and they were quick to appreciate the value of overseas branches to serve the interests of this itinerant membership.

The Institute accepted from the outset members who became shorebound in the course of promotion or change of work, but it did so with some reservation,[27] and it has retained a strong insistence on practical experience as the most acceptable training and qualification for membership, with the result that probably a smaller proportion of its members have university degrees or other academic qualifications than comparable institutions.[28] The membership did, however, overlap to some extent with other institutions, a recent estimate put members of the Naval Architects at 16-17% and another 15-20% as members of the Mechanicals.[29] Nevertheless, the Institute remains unsympathetic to attempts such as those of the Confederation of Engineering Institutions to emphasise degrees as the qualification for chartered status for engineers. At the same time it should be observed that the Institute was one of the few professional bodies to support the recommendation of the Finniston Committee on licensing or registration, as marine engineers have for long been expected to be registered by the Department of Trade (formerly the Board of Trade).[30]

In view of its comparatively practical emphasis, it is a little surprising to find the Institute with an office in the City of London. This seems to have derived almost accidentally from its origin in East London, with the attraction of being closer to its members in the Port of London rather than to Whitehall and Westminster. The first site for a purpose-built office was established in the Minories in 1913, and after the Second World War a fund in memory of members who had died in the war enabled the Institute to purchase its present site in Mark Lane, off Fenchurch Street. This location in the City has been useful in maintaining links with shipping companies and with public bodies such as the Port of London Authority.

From the beginning, the Institute has been anxious to promote regional branches for its members. Thus, an informal meeting of local members in Cardiff in September 1890 led to the establishment of the Bristol Channel Centre at Dowlais Hotel, Cardiff, the following November. The Centre showed considerable independence and enterprise, entertaining the Summer Visit of the Institution of Naval Architects in Cardiff in 1893, and the North East Coast Institution of Engineers and Shipbuilders in 1896.[31]

The *Transactions* of the Institute have appeared annually since 1889, containing the same diet of intensive specialist discussion as those of other institutions and reflecting in particular the interests of members such as the 'Battle of the Boilers' in the 1890s.[32] Members were deeply affected by the loss of the *Titanic* in 1912 and established an Engineering Staff Memorial Benevolent Fund in honour of those members who had perished in the disaster. The Institute lost heavily in both World Wars, but in December 1914 military rank was conferred on engineer officers of the Royal Navy, so it was felt that the crisis had stimulated an over-due recognition of the status of the marine engineer in the armed forces.[33] The Institute received its Royal Charter in 1933. Its Presidents have included a number of public figures in the shipping world, such as Sir Julian Foley, CB, in 1938.[34] James Adamson retained the office of Secretary until 1931, when he was succeeded by B.C. Curling, the official historian of the Institute.[35]

Of the four national professional institutions established by British engineers in the 1890s, the first was the Institution of Mining and Metallurgy (IMM), founded in 1892 and incorporated in 1915. It has stated its own objectives as being:

> 'to advance the science and practice of Mining, Mineral Technology, Mineral Exploitation and Mining Geology in respect of minerals other than coal and of Metallurgy in respect of metals other than iron; and to afford a means for facilitating the acquisition and preservation of the knowledge which pertains to the professions associated therewith.'[36]

Insofar as metal mining was an ancient industry with a long tradition of continuous activity in parts of the country such as Cornwall, the emergence of the IMM could be regarded as a development within an old industry like those which had already been established in the coal mines and iron and steel industries, particularly as it defined itself so carefully in relation to these industries. But the creation of the new institution also coincided with some very significant innovations in metal mining such as the introduction of sophisticated scientific techniques to recover metal from ores and wastes. To this extent, therefore, it can also be seen as a type of association responding to new technologies. Whatever the stimulus, however, the pattern of organisation followed that made familiar by the Civils.

The IMM was formed at a 'Preliminary Meeting' convened by a small committee, which had in turn been assembled by George A. Ferguson, the Editor of *The Mining Journal*. The meeting, attended by sixty-two men, was held at Winchester House, Old Broad Street, London, on 13 January 1892. Mr. Charles Seymour, a civil engineer, presided, and observed that no existing institution - with the honourable exception of the Mining Association and Institute of Cornwall[37] - had so far devoted itself to non-ferrous metal-mining and processing. A motion to set up such an organisation was passed unanimously, and the Institution of Civil Engineers was adopted as the model.[38] Seymour was elected President at the first General Meeting in May, and gave an inaugural address in which he expounded the civilising effects of metalliferous mining.[39] Membership grew to 615 in 1899, and to 1,799 in 1909: by 1913 it had reached 2,372.[40] Until 1954, the Institution was housed in Finsbury Circus, EC2, with offices adjacent to those of the Institution of Mining Engineers, with which it shared a library and other facilities. When the two institutions moved into separate premises the division of the library proved to be a particularly difficult operation, but with a large overseas representation amounting to about sixty per cent of the total membership, the IMM regarded it as necessary for its overseas members to maintain a full library service. The IMM received its Charter in 1915.

From the outset, the IMM took a strong stand on the professional status of its members. This was in part a reaction to the hitherto dominant tradition amongst metal miners of the practical amateur - the intrepid prospector and casual treasure-seeker who had figured prominently in the great gold rushes of the nineteenth century. By the end of the century, the last of these gold rushes were in full swing in South Africa and Western Australia, but in most of the established metal-mining fields a much more systematic and scientific attitude had been introduced into metal working as the ores became harder to reach and required more skill in extracting the metals from them. This was particularly the case in America, which was taking a pronounced lead in scientific metallurgy in the 1880s and 1890s, and the insistence on increasing professionalism amongst British practitioners was a recognition of this fact. In his Presidential Address of 1922, Mr. S. J. Speak expressed this objective succinctly:

'One of the objects of our Institution has been to make clear the distinction between the educated engineer and the so-called practical one, and this has been so far accomplished and become effective that the latter type is rapidly becoming extinct.'[41]

Coupled with this concern about their own professionalism, members of the IMM applied themselves to the acquisition of adequate educational qualifications. Education was desirable, but it was difficult to lay down standards for a membership which was territorially so scattered. Professor A. K. Huntingdon, of Kings College London, a founder-member and early President, had observed;

'... the superior advantages which American mining engineers and metallurgists enjoyed in an educational way over young Englishmen, from the opportunities afforded them of getting the best experience at home.'[42]

The Institution placed great importance on technical qualifications obtained at universities, and especially at Imperial College London, which some members considered should become a technological university for the Empire.[43]

The IMM also gave more emphasis than was usual amongst engineering institutions to the need for strict discipline in order to maintain professional standards. It was realised that members were placed in a position which was exceptionally vulnerable to temptation, through the manipulation of shares and advice based upon expert knowledge of probable mineral resources, and the annual reports contain a remarkably high number of cases of disciplinary action. In 1914, for example, the Council considered four 'alleged cases of unprofessional conduct by members', of which two were dismissed, one was proved to involve 'the wrongful use' of the initials of another technical society', and the other was 'proved in a Court of Law'. In the latter two cases, one member resigned and the other's name was 'expunged from the roll.'[44] It is not possible to be certain that this regular oversight of the conduct of members was not more evident in the IMM than in other institutions because the others were merely more discreet about their problems. But it seems likely that the Institution did have particularly severe problems associated with its determination to stamp out unprofessional metal mining and working. Walter McDermott, in his Presidential Address of 1898, had reflected:

> 'In view of the peculiar difficulties and temptations in the life of a mining engineer, I think every school of mines should have a department of ethical culture included in the regular course, with a graduated series of moral gymnastics to strengthen the conscience of the student before he is turned loose.'[45]

Whether or not his hankering after a greater facility for 'moral gymnastics' achieved any success, the expression of need is itself a recognition of a vital aspect of professional performance and a partial fulfilment of the desired objective.

In its original form, the Institution of Municipal Engineers had been the Association of Municipal and Sanitary Engineers and Surveyors, and had catered for professional engineers working on all local authority sanitary and public health projects. By the 1890s, however, some practitioners in this field had come to feel that there was a need for an organisation specialising in the sanitary aspects of these functions, and the Institute of Sanitary Engineers was duly founded in 1895, becoming the Institution of Sanitary Engineers on its incorporation in 1916, and changing its name to the Institution of Public Health Engineers in 1955. The new institution was thus essentially a break-away type, although it does not appear to have engendered undue animosity from the Municipals, and a considerable proportion of its members were also members of the Civils. It was also in part the result of a new professional *lacuna*, because after a slow start the pioneering work of Sir Joseph Bazalgette on the London sewage outfalls had been copied in towns all over the country, creating a new sort of expertise. However, it should be said on this point that few radical innovations have been introduced in methods of sewage disposal since the formation of the Institution, so that its function should rather be seen as one of maintaining certain specialist standards in public health provision. But the close association of several local Medical Officers of Health with the Institution has ensured that it has looked into wider matters of public health such as the introduction of chlorination in water supplies.[46]

The Institution attached great importance to regulating the entry into its specialised field by the introduction of qualifying examinations, which covered subjects like surveying and building methods as well as sanitary engineering. To become a Fellow, the highest grade of membership, it was necessary to have had approved practical experience in addition to these examination qualifications.[47] Membership figures are hard to determine with precision in the early years of the Institution, but by 1914 they had reached a total of 635, including nineteen Honorary, 220 Fellows, 264 Associates, and 132 Students.[48] By 1980 the total had risen to almost 3,500 with membership overseas showing particularly vigorous growth as places like Hong Kong and Malaysia have undertaken ambitious public health projects.

The Institution of Public Health Engineers seems to have produced few outstanding personalities. Its presidential addresses over the years have stressed the importance of sanitary engineering and the need to perform it professionally, but they have not demonstrated much vision of a more general nature. In view of the overall pattern of institutional proliferation, however, the views of Henry C. Adams, who gave his second Presidential Address in 1933, are of some significance as an indication of changing attitudes in the profession. He said:

'In my opinion, there are far too many separate and distinct engineering societies. They overlap in membership, and they overlap in work... For myself I am a member of eight societies at the present time...'[49]

Awareness of a growing problem, however, did not lead to any suggestions for its solution.

The provision of pure water supplies to urban communities and society at large has been a major preoccupation of engineers from antiquity to modern times, and the nineteenth century produced a distinguished tradition of men such as Telford, Simpson, Bateman and Hawkesley, who performed important engineering works in this field. So important was it, indeed, that it seemed until late in the nineteenth century that the professional interests of engineers engaged in water engineering were very adequately covered by the Institution of Civil Engineers. The formation of a new organisation for this purpose must, therefore, be regarded as a break-away from the Civils, although as with the Sanitary Engineers and other specialist groups they seem to have departed from the parent association without grudges on either side, and with most members of the new body retaining their membership of the Civils. The sympathetic support of the Civils seems to have become a regular feature of these divisions, and suggests that the older organisation saw the new one as akin to what would later have been regarded as a specialist 'section'. Institutional proliferation was becoming something of a habit, and nobody seemed to object.

A preliminary meeting to establish the new organisation for water engineers was held in Nottingham on 11 January 1896, with thirty-three engineers present and with twice that number having signified their willingness to support an association. D. V. F. Gaskin, Water Engineer to Nottingham Corporation, had taken the initiative in summoning the meeting, and was elected the first President:

'He wished it to be understood that they were not departing from the Parent Institution in Great George Street, but that that Institution could not take up many things that they, as an Institution, would take up.'[50]

The first general meeting was held on 11 April 1896, and the title 'The British Association of Waterworks Engineers' was adopted. At the end of 1896 there were 110 members and thirty-two associates on the roll. The word 'British' was omitted from the title in 1906, and in 1911 the organisation was incorporated under the title 'The Institution of Water Engineers'. Membership then stood at 396, and by 1914 it had risen to 422.[51] The Institution thus settled down quickly to perform a familiar pattern of professional functions in the highly specialised field of its members.

Problems of heating and ventilating buildings became increasingly pressing in the nineteenth century as factories, offices, and shops grew in size, and those engineers who became involved in the provision of such services had begun to think of themselves as a distinct professional group by the end of the century. In 1898 the Institution of Heating and Ventilating Engineers was formed. Its membership began around a hundred and rose to 230 by 1904. By 1908 it had reached 280, but Mr William Yates, the President in 1909, regarded the situation of the last few years as being 'almost stationary' and said that it 'cannot he considered satisfactory'.[52] There was then an encouraging growth up to 476 in 1914.[53]

The constitution of the Institution, adopted in 1898, expressed its objectives as being:

'to promote the intellectual welfare of members, to read, consider and discuss papers on problems of Heating, Ventilation, or other kindred subjects and to take measures to extend, develop, or safeguard the interests of these trades.[54]

The emphasis on 'intellectual welfare' was apparent from the outset, with frequent discussions on educational provisions and examination requirements. A useful link was made with University College London in 1909, when a section on heating and ventilating was introduced into the engineering course there, although this does not appear to have been a success as there was no mention of it in the records of the Institution after 1912.[55] The City and Guilds introduced an examination in heating and ventilating engineering in 1911, and in 1920 the Institution introduced its own examinations, confirming thus the professional identity of its own specialist skills. Concluding his Presidential Address in 1950, when the membership of the Institution stood at around 1,800, Mr. R. Duncan Wallace was able to review half a century of successful development and to claim with justification that:

'An Institution such as ours confers professional status on its members... I would say most definitely that heating and ventilating engineering is a separate science and should be treated as such alongside civil engineering, mechanical engineering, electrical engineering, architecture and physics, and in recognising this fact, your Council (has) petitioned His Majesty for a Charter of Incorporation'.[56]

The Institution had certainly succeeded in establishing itself as a vigorously independent specialisation in the spectrum of modern engineering.

The dawn of the twentieth century saw the proliferation of professional engineering institutions continuing unabated, with four new institutions between 1900 and 1914. The first of these was the Institution of Automobile Engineers. In retrospect, this can be regarded as a break-away from the Mechanicals, because in 1945 it amalgamated with the Institution of Mechanical Engineers and formed a specialist section in that organisation. Such an interpretation would be slightly misleading, however, because automobile engineering could be regarded as a distinctively novel type of engineering at the beginning of the century, and the new Institution emerged to cater for the specialist interests of engineers in this field. As Mr. H. G. Burford said in his Presidential Address to the Institution in 1924, 'It can be said quite safely that no Industry in the world has grown so rapidly as that of automobile engineering'.[57] The internal-combustion engine in all its forms, but most particularly in the shape of the automobile engine, has transformed the society of advanced industrial nations in the twentieth century, and the roots of this development were set in the closing decades of the previous century, when the motor-car appeared as a viable means of personal transport. A considerable manufacturing industry had already been established to produce automobiles by the beginning of the twentieth century, and this expanded with spectacular speed before and after the First World War. The industry in Britain had become sufficiently conscious of its on importance to promote the formation of the Institution of Automobile Engineers in 1906.

Beginning with under 200 members, the new Institution had already reached 900 by 1914, and by 1939 it had passed 3,300 members. Its development followed the well-defined lines of institutional proliferation into new fields of engineering expertise which had not received sufficient attention from existing organisations. Following a statement of objects which mirrored the aspirations of other institutions in stressing the value of a greater sharing of knowledge, it drew up a constitution which resembled that of older-established bodies in virtually every detail, including the hierarchy of grades of membership. It encouraged the education of its members by stipulating examination qualifications in addition to working practice. It discussed matters of keen interest to the industry such as the standardisation of parts and the condition of roads. And it worked towards the achievement of corporate status by Royal Charter, which came in 1938.[58]

Perusal of successive presidential addresses suggests that the Automobile Engineers were amongst the first to reflect a change in tone from that of previous institutional practice in this office. Engineers had always been anxious about their status, and the establishment of the professional institutions had done much to clarify the principles of employment and salary structures without ever engaging directly in the negotiation of such matters. But in the speeches of IAE Presidents after the First World War there appeared an increasing stridency in comments on these subjects, arising from a consciousness of an almost indefinable loss. As Professor W. Morgan pointed out in 1930:

'The engineer occupies a position inferior in privilege to that of members of, say, the legal and medical professions... Clearly improvement in the position of the engineer must be preceded by public appreciation of the nature of his services and the arduous character of his work.'[59]

It is unlikely that any of the great nineteenth century engineers would have felt so strongly this imposition of inferiority, and it probably reflects the consciousness of engineers in the twentieth century of the increasing diffuseness of their profession. There were many more professional engineers, but they were more widely scattered over more industries and specialist activities than in the mid-nineteenth century, and as a result they began to feel vulnerable to unfavourable comparisons with other professions. They even became anxious about the definition of their own profession. Mr. H. Keir Thomas, in his Presidential Address to the Automobile Engineers in 1926, had applied himself to this problem, going back to Vitruvius to establish the antiquity of engineering and considering various other attempts to give some precision to the concept of the professional engineer. He cited what he called the 'American definition' as adopted by the American Society of Mechanical Engineers:

> 'Engineering is the science of controlling the forces and utilising the materials of nature for the benefit of man, and is the art of directing and organising human activities in connection therewith.'[60]

Curiously, the speaker showed no signs of recognising that this had been modelled on Tredgold's famous definition framed for the Charter of the Civils in 1828. The 'American definition', like the 'American method' of manufacture by interchangeable parts, had clear British precedents. The inadequacy of historical appraisal is of some significance here, for it reflects the increasing incomprehension of twentieth century engineers about what was happening to their own profession.

Like the Automobile Engineers, the Institution of Locomotive Engineers could be regarded as a break-away from the Mechanicals in so far as it eventually merged with the older body in 1969, but unlike the Automobile Engineers its preoccupation with railway locomotives fell squarely within the original sphere of interest of the Mechanicals, so that the degree of deliberate divergence was more pronounced. However, the Locomotive engineers could also be regarded as the product of various local societies, and of more general enthusiasm for railway locomotives - particularly steam locomotives - than just amongst engineers engaged professionally with them, so that it is not possible to categorise it precisely.

By the end of the nineteenth century, locomotive engineers and related enthusiasts were beginning to feel that their interests were being neglected by the national institutions. Thus in 1893 the GWR Engineering Society was formed at Swindon and received encouraging support from the Directors and Locomotive Department there.[61] Similar societies sprang up elsewhere, and several publications on railways and locomotives appeared, which led to the formation of a Railway Club in 1899. This, however, was felt to be of too general an appeal to some members, who broke away to form the Stephenson Society in 1909. The leading members of this were two friends, L. E. Brailsford, a Banker of Croydon, who became Chairman, and G. F. Burtt, of the London Brighton and South Coast Railway Works in Brighton, who became Secretary. But the two friends proceeded to fall out, polarising a division of interests between the 'amateurs', who remained with Brailsford in the Stephenson Society, and the 'professionals', who left with Burtt to form the 'Junior Institution of Locomotive Engineers', which was inaugurated in 1911 and promptly dropped the prefix 'Junior' from its title. Burtt became Treasurer, and at the end of its first year it numbered fifty-two members.[62] Many of these early members were Brighton men, but the Institution quickly established itself as a national body and came to number some of the most distinguished locomotive men of the twentieth century such as Gresley, Stanier and Bulleid, amongst its officers. Its membership grew to

178 in 1915, when it was incorporated, and had reached 485 by 1919.[63] The President of the Locomotive Engineers in 1927-28 was H. N. Gresley (he became Sir Nigel Gresley in 1936), then Chief Mechanical Engineer of the London and North Eastern Railway. Being also a member of the mechanicals, Gresley suggested that it would be mutually advantageous if the Locomotive Engineers could be assimilated as a specialised section into the Mechanicals, but the Mechanicals' qualifying examination was regarded as unacceptable by many of the Locomotive Engineers, so that nothing was done about it then. However, in 1931 the Locomotive Engineers transferred their meeting place to the hall of the Mechanicals, and in 1969 the reconciliation envisaged by Gresley was finally accomplished, the 2,260 members of the Locomotive Engineers then forming the Locomotives Section of the Institution of Mechanical Engineers.[64]

Aeronautical engineering was another novel area of specialisation which achieved identity and institutional form in the twentieth century. The Royal Aeronautical Society had been formed as early as 1866 as the Aeronautical Society of Great Britain, but for several decades its membership had been confined to less than a hundred enthusiasts for aeronautical experiments. It had begun to grow rapidly in the first decade of the new century as heaver-than-air flight became established, attracting engineers interested in both the aerodynamics of flight and the special problems of aero engines. In 1911 the Society was reorganised with new grades of Fellow and Associate, demonstrating that it had begun to assume the characteristic features of an engineering professional institution. Membership had reached 348 in 1912. In 1918, as a result of its important contribution to the war effort in the First World War, the Aeronautical Society acquired the unusual honour of the prefix 'Royal'. It had another unusual distinction: in 1898 women had become eligible for full membership, which made it unique amongst British engineering institutions, the profession otherwise retaining a virtually complete male monopoly until after the First World War. At its eightieth anniversary celebrations in 1946, the Society had some 5,000 members organised in twenty-four branches.[65]

The fourth and most substantial of the institutions which were formed or came to prominence in the early century was the Institution of Structural Engineers, which was founded in 1908. It was originally established as 'The Concrete Institute', because it was felt by some prominent civil engineers that the new structural material - concrete - was not receiving the engineering and scientific attention which it deserved. In particular, the Civils had not devoted much time to considering the implications of concrete for engineering. Other bodies, including the Royal Institute of British Architects, the War Office, and the British Fire Prevention Committee, expressed some interest in the subject, and it was the Chairman of the latter, Edwin O. Sachs, who convened the first meeting of the Council of the Institute in the Ritz Hotel, Piccadilly, on 21 July 1908. The Earl of Plymouth took the chair on this occasion, and Sachs was appointed Chairman of Executive. The familiar process of institutional evolution followed rapidly thereafter, with the Institute securing incorporation under the Companies Act in 1909, and with a constitution and hierarchy of membership grades being set up. There were about a hundred founder members, and at the end of the first year membership had risen to 485. The total had increased to 1,006 in 1913, and it continued to grow to become one of the leading engineering institutions with over 14,000 members in 1980.[66]

The Concrete Institute transformed itself into the Institution of Structural Engineers on 28 September 1922, and went on to receive its Royal Charter in 1934. Britain is the only nation to have a separate institution for structural engineering: elsewhere engineers in this field have remained part of civil engineering. The Structurals enjoy a substantial overlap of membership with the Civils, and also with the Mechanicals and Highway Engineers, but they have nevertheless preserved a vigorous independence of other bodies. They are the only major British engineering institution to insist on a written examination for corporate membership. They have developed important overseas connections, so that flourishing branches now exist in Hong Kong and elsewhere, and they have supported attempts to improve consultation and collaboration between engineering institutions. The Gold Medal of the Structurals has been awarded to a series of distinguished engineers, including Ove Arup (1973), Oleg Kerensky (1977), and Professor Alec Skempton (1982). Unlike most engineering institutions, the Structurals have been well-served by historically-minded members such as Lt. Col. C. H. Fox, who compiled an historical survey of the Institution in the 1930s, and Dr. S. B. Hamilton, who wrote an account of its history for the Jubilee celebrations in 1958.[67]

This outline of the process of institutional proliferation in the British engineering profession down to 1914 has dealt only with national organisations. Regional and local associations will be surveyed separately in the next chapter, but even amongst the national organisations there were several which have not been mentioned as being too small or as being too peripheral to the main stream of the engineering profession. The Society of Engineers, for example, has been more of a social body than a professional organisation, although many professional engineers have been members of it since it was founded in 1857. This derived from a pioneering educational institution, the College for Civil Engineers, which had been established at Putney in 1834. A group of old students of the College formed the Putney Club, which then became the Society of Civil Engineers, and subsequently amalgamated with the provincial Civil and Mechanical Engineers' Society. It was incorporated in 1910, and its membership grew from 208 in 1861 to 500 in 1899 and to 3,416 in 1963.[68] Similarly, the Institute of Metals is on the borderline of our study: founded in 1908 with 355 members, it grew to 645 in 1914 and to 8,947 in 1970, but it has not operated strictly as an engineering professional institution.[69]

This survey has inevitably been somewhat repetitive, because each new development of a national institution has tended to take the same form and to pursue similar patterns of evolution. The uniformity of response by a sequence of specialist groups to the consciousness that their interests have not been sufficiently well represented by existing institutions has, after all, been the most significant feature of the whole process, even though the more interesting features have usually been those where the quality of personal leadership or the novelty of the situation has imposed some characteristics which have varied this pattern. Viewed from the point of view of the participants, the decision to establish a new institution must have seemed the obvious solution to the problems which they encountered. But in a more general and longer perspective, the most important questions are: Need it have happened? Was the process of institutional proliferation really necessary? And did the engineering profession in Britain lose more than it gained from the fragmentation which occurred between the mid-nineteenth century and 1914?

There are no simple answers to these questions. Clearly, there were conceivable alternatives to the process of institutional proliferation, particularly the possibility of the expansion of the profession being contained within a single association characterised by some sort of decentralised or sectional structure. If such a unified organisation had been sustained, there can be little doubt that the overall social status of the profession in the twentieth century would have been greatly enhanced. But in the circumstances of British society in the nineteenth century, with its trenchant individualism and insistence on the ethos of self-help, it would have been unrealistic to have attempted to maintain such a monolithic organisation. The engineering profession responded to the conditions and opportunities of the times, and within these parameters it achieved much. To have expected more of it would have been to have expected too much. Proliferation and division were thus part of the unavoidable legacy of British industrialisation in the nineteenth century, and it is pointless to deplore them and their consequences. It is not unreasonable, however, to use the historian's gift of hindsight in order to diagnose the troubles of the British engineering profession in the twentieth century as attributable, in large measure, to a pattern of development in the nineteenth century which was ultimately divisive, in so far as it encouraged engineers to magnify their specific skills at the expense of the greater unity of the whole profession.

Notes

1. See Chapter 9 for a discussion of this Report.
2. 'Report of the Court of Inquiry... upon the circumstances attending the fall of a portion of the Tay Bridge on the 28th December 1879', *P.P. 1880* (c.2616) xxxix, the report of Mr. Rothery, p.44, paragraph 120.
3. For a fuller discussion of these developments, see Chapter 9.
4. The validity of the term is discussed by S. B. Saul, *The Myth of the Great Depression, 1873-1896*, London, 1969.
5. R. Appleyard, *The History of the Institution of Electrical Engineers (1971-1931)*, 1939, p.29, quoting Minutes of Council.
6. Ibid., p.35.
7. Ibid., quoted p.20, but the passage is not verbatim with the *Minute Book* which has 'engineers' for 'enquirers'.
8. Ibid., quoted p.22.
9. L. T. C. Rolt, *Victorian Engineering*, 1970, p.214, tells the story of the telegraph being used to secure the arrest of John Tawell, who had escaped on a train from Slough to Paddington after poisoning his mistress.
10. Appleyard, *op. cit.*, 26.
11. Ibid., p.29-30, where I. K. Brunel is wrongly described as the President of the Civils at the time.
12. Ibid., p.37.
13. Ibid., p.38.
14. Ibid., p.43.
15. Ibid., p.44.
16. *Minutes of Council IEE*, 19 November 1879.
17. Ibid., 10 November 1887.
18. Ibid., 25 October 1888.
19. Appleyard, *op. cit.*, pp.47-48.

20. Ibid., 51: see also Chapter 9 for a discussion of the role of Ayrton and other academic engineers.

21. Ibid., p.6.

22. Ibid., quoted p.32. It should be observed, however, that Hertha Ayrton, the wife of Professor W. E. Ayrton, was elected a Member in 1899 - the only woman to be so elected until 1916: see Appleyard, *op. cit.*, pp. 167-8.

23. B.C.Curling, *History of the Institute of Marine Engineers*, 1961, p.2.

24. Ibid., quoted p.3.

25. *Trans. Inst. Marine Engs.*, 2, 1891, 8; 12, 1901, 9; 26, 1915, ix

26. See Chapter 5, note 25, for this distinction.

27. Curling, *op. cit.* p.5.

28. Verbal communication with the Secretary, 1 February 1980.

29. Ibid.

30. Ibid.

31. Curling, *op. cit.*, p.201 and p.202.

32. Ibid., p.202: the issues involved the safest and most efficient boilers for generating high pressure steam.

33. Ibid., p.207. However, there were further difficulties over this vexed issue: see R. H. Parsons, *A History of the Institution of Mechanical Engineers*, 1947, p63-64.

34. His Presidential Address is in *Trans. Inst. Marine Engs.*, 50, 9, 1938-39, 99-202.

35. Curling retired in 1951, to be succeeded by the present Director and Secretary, Mr. J. Stuart Robinson.

36. IMM, 'Objectives and Activities', April 1980.

37. See below, Chapter 7.

38. Cited in Ferguson's letter, as quoted in *Trans. IMM*, 1, 1892-93, 34, in the report on the preliminary meeting.

39. *Trans. IMM*, 1, 1892-93, 50-71.

40. Ibid., reports for years cited.

41. S. J. Speak, 'Presidential Address' in *Trans. IMM*, 31, 1921-22, xxx.

42. *Trans. IMM*, 6, 1897-98, 183, where he is quoted indirectly by McDermott.

43. Speak, *op. cit.*, note 41 above, p.xxxiv.

44. *Trans. IMM*, 23, 1914-15, xiv.

45. Ibid., 6, 1897-98, 192.

46. See, for example, E. A. Whitlock's paper on this subject in *Jnl. Inst. Sanitary Engs.*, 53, 1954, 61.

47. Verbal communication from the Secretary, 8 May 1981.

48. *Jnl. Inst. Sanitary Engs.*, 18 February 1914.

49. Ibid., 37, 1933, 4.

50. Quoted in the 1936 Presidential Address of Mr. N. J. Peters, *Trans. Inst. Water Engs.*, 41, 1936, 16-17.

51. *Trans. Inst. Water Engs.*, 19, 1914, 8.

52. *Proc. Inst. HVE*, 10, 1909, 19.

53. Ibid., 14, 1914.

54. Quoted in P. L. Martin, 'Noblesse Oblige' in *Jnl. Inst. HVE*, 39, May 1971, 25.

55. Martin, *op. cit.* 28.

56. *Jnl. Inst. HVE*, 18, 1950-51, 17.

57. *Proc. Inst. Auto. Engs.*, 18, 2, 1923-24, 100.

58. For the subsequent amalgamation negotiations with the Institution of Mechanical Engineers, see Parsons, *op. cit.*, p84-86.

59. *Proc. Inst. Auto. Engs.*, 24, 1929-30, 5-7.

60. Ibid., 20, 1925-26, 9 and 11.

61. H. Holcroft, 'History', in *Jnl. Loco. Engs.*, 50, 1960-61, 663.

62. See the 1956 'Presidential Address' by J. F. B. Vidal in *Jnl. Loco. Engs.*, 46, 1956, 296-306, which gives more details than Holcroft on several points.

63. Holcroft, *op. cit.*, p.666.

64. *Jnl. Loco. Engs.*, 59, 1969-70, 131.

65. For the history of the Society, see *The Royal Aeronautical Society 1866-1966: A Short History*, published by the Society, 1966.

66. *Trans. Concrete Inst.*, 1, 1, 1909, v; 2, 1910, 72; and 5, 1913, 406. Also see the quantitative review of membership in Stanley Vaughan's Presidential Address, 'Twenty-one years' Progress as a Chartered Institution', in *The Structural Engineer*, 33, December 1955, 365-375.

67. Lt. Col. C. H. Fox, 'The History and Progress of the Institution of Structural Engineers', in *The Structural Engineer*, Royal Charter Issue, 13, new series, 1935. Dr. S. B. Hamilton, 'The History of the Institution of Structural Engineers', in *The Structural Engineer*, Jubilee Issue, July 1958, 16-21.

68. See W. H. G. Armytage, *A Social History of Engineering*, 1961, p150-1. Also *Trans. Soc. of Engineers*, 1 of which appeared in 1861. For the figures cited, see *Jnl. and Trans. Soc. of Engs.*, 1963, Report to Council, 235.

69. See, for instance, the *Jnl. Institute of Metals*, 78, 1950-51, 668, where there is a graph summarising the development of this body in the 'Report to Council'.

Chapter Seven

Regional Diversification*

In parallel with the proliferation of specialised national engineering institutions which has been examined so far, there also occurred, in the second half of the nineteenth century, a remarkable flowering of regional activity in the British engineering profession, which produced its own crop of vigorous institutions. These organisations were quite separate from the national institutions, and generally differed from them in stressing territorial rather than specialist engineering qualities as conditions of membership, although all of them were strongly influenced by the types of engineering which predominated in their regions. We have already observed an element of ambiguity in the origin of the Institution of Mechanical Engineers which, with the strong Birmingham and provincial emphasis of its early years, could be regarded in some respects as a regional institution. The subsequent development of the Mechanicals, however, makes it appropriate to consider it as a national organisation from the outset. Another development, even earlier than the Mechanicals, requires special consideration before we proceed to survey the regional institutions. This was the formation of an institution of professional engineers in Ireland.

The first replication of the Institution of Civil Engineers occurred in Dublin in 1835 with the establishment of the Civil Engineers' Society of Ireland, better known by the title which it adopted in 1844, the Institution of Civil Engineers of Ireland.[1] The moving spirit and first President was Colonel John Fox Burgoyne (1782-1871), later Baronet (1856) and Field Marshal (1868). Burgoyne was a member of the Corps of Royal Engineers and a veteran of the Napoleonic Wars. In 1831 he was appointed Chairman of the newly constituted Board of Public Works, Ireland,[2] formed to promote inland navigation, fisheries, roads, bridges, public buildings and harbours, and as such he served to 1845 as 'a sort of engineering proconsul'.[3] Addressing the twenty members who attended the first meeting of the Civil Engineers' Society in 1835, Burgoyne deplored the 'low ebb' of the profession in Ireland. He justified the formation of the Society by distinguishing three different processes for acquiring knowledge of civil engineering: by practical experience, by study, and by 'personal intercourse and mutual communication between members of the same profession'.[4] Thus from the outset the founder identified the mutual educational function of the professional institution as the role which the Society should seek to fill for Irish engineers.

The support of Burgoyne and other senior Royal Engineers was crucial to the new organisation, and when they were not able to devote much time to it, the Society languished. After enrolling forty-two members and meeting for two years it lapsed into inactivity, probably because Burgoyne and his colleagues were preoccupied with major drainage

---

* See Table II on page 234, *Regional Professional Organisations of Engineers in Britain*

schemes and harbour works, as well as with projects to build railways in Ireland. Burgoyne was one of the Railway Commissioners appointed in 1837 to plan railway developments in Ireland and to survey possible routes: C. B. Vignoles and J. MacNeill were engaged to make preliminary surveys in the south and north of the country respectively, and by 1840 the construction of Irish main lines was well under way.[5] It was not, therefore, any lack of engineering activity which caused the professional organisation to make such a faltering start, yet the prospect had come to look so unpromising by 1844 that Burgoyne seriously considered winding up the Society and dividing its assets between its members, but before doing so he consulted Robert Mallet, a young engineer who was a partner in a family foundry in Dublin which was flourishing through business provided by the railways, and the result was that the organisation survived its crisis.

Mallet quickly diagnosed the trouble to be that the Society had been 'founded upon too narrow a basis - confined to civil engineers in the older and narrowest acceptation of the title', and that it had only appointed honorary officers.[6] He recommended remodelling the bye-laws even more closely upon those of the London institution, and appointing a salaried secretary. These proposals were promptly adopted. The reconstituted Institution met in October, with its new title and new rules, and with Thomas Oldham as its first paid Secretary, and it has continued to meet ever since. The first volume of its *Transactions* appeared in 1845.[7]

Burgoyne returned to England in 1845, but he was succeeded as President by a series of eminent engineers, some of them being men like Vignoles with international reputations.[8] The period of the Irish Famine in the late 1840s was one of considerable public investment in railways and drainage works, so that there was plenty of work for engineers. Railway building, in particular, went on apace. This activity fell off somewhat in the 1850s and 1860s, due to the increased reluctance of the British Government to provide public resources. Mallet, whose own business was so severely affected that he moved his base to London, used his Presidential Address in 1866 to deplore the introduction of *laissez faire* doctrine into Ireland:

> 'I therefore venture to record my opinion, that it was a disastrous policy, which reduced the Functions of the Board of Public Works in Ireland almost to a cipher, and opposed to which the sonorous nonsense of Anglo-Saxon enterprise and the wisdom of self reliance are no refutation.[9]

This was an unusually strong statement of an unorthodox political opinion for the President of an engineering institution, but Mallet was doubtless correct in identifying a negative attitude on the part of central government to the need for industrial and economic development in Ireland. 'The social circumstances of Ireland are exceptional', he concluded, 'and peculiarly unfavourable in some respects for the development of engineering or industrial works'.[10]

Nevertheless, the Institution of Civil Engineers of Ireland survived. It declined from a total of 107 members in 1847 to eighty-eight in 1869, but it recovered sufficiently to carry through a campaign to achieve a charter of incorporation in 1877 and by 1889 membership had risen to 192. By 1909 it had reached 327.[11] Quite apart from the obvious difference of topography and personalities, the Institution differed from its counterparts in mainland Britain in several important respects. For one thing, the peculiar semi-colonial status of Ireland made engineering activity unusually dependent on public works and the talents of Royal Engineers, at least in the early years of the organisation. Even more important, however, was the fact that the Irish Institution survived to become the leading national professional engineering organisation of an independent state, so that some of the centripetal factors which worked strongly in the twentieth century to undermine regional organisation in Britain did not affect it. For this reason, if for no other, it requires treatment as a separate case and should not be regarded as a typical product of regional diversification, despite the historical fact that it was the first regional society to seek deliberately to copy the model of the Civils.[12]

If, then, we exclude the Dublin Institution as well as the Mechanicals, the first regional society was the North of England Institute of Mining and Mechanical Engineers, founded at Newcastle-upon-Tyne in 1852. By the time that the Sheffield Society of Engineers and Metallurgists had been formed in 1894, at least a dozen of these regional organisations were flourishing for which some documentary evidence has been recovered. It is not suggested that this list exhausts the material: indeed, there is clear evidence of the existence of several other associations of engineers,[13] but there is sufficient material here to provide a fair reconstruction of the pattern of the regional diversification of the British engineering profession.

Instead of pursuing a topographical or chronological survey of these institutions, the links with local business enterprises are more clearly revealed if they are grouped according to which type of engineering predominated in the formation of each institution. On this basis, they fall into three groups, each of which tended to produce a characteristic organisation, with distinctive aims and procedures. These are, first, the institutions promoted primarily by mining engineers; second, those in which mechanical engineers were most influential; and third, institutions formed by civil engineers and shipbuilders. By analysing them according to these categories, it can be shown how they arose out of local resources to meet needs which varied in detail between different types of organisation as well as between different parts of the country. All types of regional professional institution, however, were subjected to the same centripetal forces in the twentieth century which caused them to contract and, in many cases, to disappear. The institutions showed varied capacities for meeting this challenge to their continued existence, and the causes of the process of differential collapse will be examined at the end of the chapter.

The North of England Institute of Mining and Mechanical Engineers was the first significant regional association of engineers. Founded in 1852, it was incorporated by Royal Charter in 1876, and it survives today in Neville Hall, the neo-Gothic office block built to accommodate it in Newcastle-upon-Tyne in 1872. The Institute began as one of coal mining engineers, the phrase 'and Mechanical' being added to the title in 1866, and the emphasis on coal mining has remained strong down to the present day. Its founder was Nicholas Wood, a mining engineer with an early interest in railways, being a friend of George Stephenson and a judge at the 'Rainhill Trials' in 1829. Wood became the first President of the Institute and held the post until his death in 1865. The dominant anxiety which called the Institute into existence was the alarming frequency of bad accidents in coal mines, to combat which engineers, coal mine owners, and local residents were prepared to join together, in the words of the initial resolution:

'to meet at fixed periods and discuss the means for the ventilation of coal mines for the prevention of accidents, and for general purposes connected with the winning and working of collieries.'[14]

About eighty members were enrolled in the first year, and by 1865 membership had risen to over 300. By 1876, when it received its Charter, there were over 800 members. It is impossible to tell from the lists what proportion of these were mining engineers, but it was certainly large, reflecting the dominant industry of the Northumberland and Durham coalfield.

The Institute gave prolonged attention to the problems of mine ventilation and lighting, and stimulated a succession of technical papers which made an important contribution to the solution of these problems. It also associated itself with attempts to improve the quality of technical instruction in the coalmining industry. As early as January 1853 the Council of the Institute was canvassing the idea of a School of Mines in Newcastle, and two years later a detailed proposal for such a venture was prepared for the General Meeting.[15] After another two years of raising support and money a Bill was drafted for submission to Parliament.[16] Thereafter, however, the educational effort was channelled into support for a School of Mines at the University of Durham, and for the foundation of the College of Physical Science at Newcastle which was achieved in 1871. This body subsequently became King's College and the core of the University of Newcastle.

Despite its strong roots in the North East, the Institute was willing to co-operate with any other organisation to achieve greater safety in coal mines, and established national and even international links with this object in view.[17] The same concern led to its promotion of the Federated Institution of Mining Engineers which was formed as a national organisation in 1889 and became the Institution of Mining Engineers.[18] The President of the North of England Institute, Mr. John Daglish, had urged the establishment of such a Federation in his Presidential Address of 1886, and the Secretary, Mr. T.W. Bunning, had argued in a similar vein the following year.[19] Consultations followed with half a dozen other regional mining institutions, as a result of which a meeting was called in Sheffield in January 1889 at which the Federal Institution was inaugurated. Although the regional organisations preserved their identity within the Federation, the process of gestation demonstrates the unusually strong connection between some regional and national institutional organisations, which was made possible by the pronounced decentralisation of the British coal mining industry.

By campaigning for a national organisation of mining engineers, the North of England Institute deprived itself of part of its *raison d'etre* once this objective had been achieved, and the membership consequently declined from above 800 in 1876 to about 540 in 1981.[20] The Institute has suffered a further narrowing of its functions with the expansion of the state system of technical education and with the nationalisation of the coal mining industry. As a result of this contraction, Neville Hall is now let out in part to other bodies. Nevertheless, the Institute survives as a venerable organisation with many significant achievements to its credit.

Another important regional association of mining engineers was the South Wales Institute of Engineers. The initiative for the formation of this organisation came from the iron industry, in the person of William Menelaus, the General Manager of Dowlais Ironworks, at that time the largest ironworks in the world,[21] and the strong representation of the iron industry was demonstrated by the choice of a blast furnace as the emblem of the Institute when it was established at Merthyr Tydfil on 29 October 1857. The main object was clearly stated:

'The South Wales Institute of Engineers shall devote itself to the encouragement and advancement of Engineering Science and Practice; it being established to facilitate the exchange of information and ideas amongst its members, and to place on record the results of experience (elicited in discussion).'[22]

This general educational purpose was stressed by Menelaus in his inaugural address, which was devoted to the means of achieving the elimination of waste by mechanisation and by improved management and scientific application:

'If education is not to be the means of lightening toil and lessening the hours of work, small good will it do.'[23]

Education, in short, was presented as a valuable but strictly functional process for improving industrial performance.

By the time that the Institute applied successfully for incorporation by Royal Charter in 1881, the dominance of the iron industry in its membership had apparently been replaced by that of the coal mining industry, reflecting regional changes in the relative importance of the two industries. The headquarters had been moved from Merthyr to Cardiff in 1878, and the Institute became increasingly preoccupied with matters of coal mine engineering. A new office building was constructed on a site in Park Place in 1894, 'in red terracotta and red pressed factory bricks as was the vogue at that time'.[24] The building has survived the implied aesthetic censure of these remarks, and although grimly dilapidated and largely occupied by the University of Wales Institute of Technology, it continues to provide essential services for the Institute, with the blast furnace 'logo' still picked out in coloured tiles on the floor of the entrance hall, and with various items of Institute furniture on display. At its centenary in 1957, the Institute had a total membership of 911: it had been falling from around 900 in the 1920s, but had been boosted to 989 by its affiliation to the Institution of Mining Engineers in 1952. This affiliation had probably been delayed by the continued association of members from the iron and other industries, but its achievement indicated that the Institute had become an organisation of coal mining engineers. However, the process of gradual decline in membership resumed thereafter, as the South Wales coal field contracted in scope. In its centenary year, the Institute had a healthy Benevolent Fund with a capital of £16,500, with which it was claimed that: 'it has been able to help distressed members or their dependents on numerous occasions.'[25] Although there have been some prominent industrialists and coal owners amongst its members, the Institute has only partially fulfilled the requirements of a professional association of engineers. Yet despite quarrels in South Wales regarding the provision of training in mining techniques which produced rival teaching establishments,[26] there was no viable alternative to it until the formation of local branches by the national institutions.

Chesterfield provided the original home for another regional engineering institution which relied heavily on recruitment from the coal mining industry. The Chesterfield and Derbyshire Institute of Mining, Civil and Mechanical Engineers was founded in 1871:

> 'It originated from the desire of engineers of the Chesterfield industrial region to meet and discuss mutual problems which were met with in the coal mines, iron works, blast furnaces, gas works and foundries under their control.'[27]

Professor Hinsley, Emeritus Professor of Mining Engineering at the University of Nottingham and an active member of the Institute, has written a memoir of the association in which he records that it was supported by: 'nearly all the heads of the large firms around Chesterfield', although interestingly one who dissented in 1871 was Charles Markham, Managing Director of Staveley Coal and Iron Company, who considered that such bodies had already become too numerous.[28] The Institute quickly acquired over a hundred members, and Lord Edward Cavendish, a younger son of the Duke of Devonshire, accepted the office of first President. Early papers were devoted to the controversy about mine ventilation systems between supporters of the rival Waddle and Guibal fans,[29] and the Institute undertook the construction of a 'Stephenson Memorial Hall' in Chesterfield, a few miles from George Stephenson's last home at Tapton House. The title of the Institute was changed in 1886 to 'The Chesterfield and Midland Counties Institution of Engineers' and in 1889 it became a founder-member of the Federated Institution of Mining Engineers so that, like the North of England Institute in Newcastle, it acquired in part the character of a national institution. It had supported the initiation of a course in mining at the University College of Nottingham in 1888, and from 1890 onwards the majority of its meetings were at the college, this change in venue being reflected in a further modification of the title to 'The Midland Counties Institution of Engineers' in 1901.[30] Membership reached 375 in 1901, and then remained fairly steady until nationalisation of the coal mines after the Second World War, when it rose to a peak of 859 in 1965, but then fell to 598 in 1973, the last date covered in the Centenary study.[31] The Institution appears to have fulfilled the modest regional requirements of professional mining engineers, but it seems probable that, like other provincial organisations of coal mining engineers, it has derived much of its support in the twentieth century from the national institution with which it has been federated, and for which it has acted, in effect, as a local branch.

Although coal mining was responsible for the largest body of British mining engineers in the nineteenth century, the traditional metal mining industries of Cornwall and Devonshire also witnessed some vigorous regional activity of professional engineers. Nationally, this group was represented by the Institution of Mining and Metallurgy, established in 1892,[32] which operated independently of the coal-mining dominated Institution of Mining Engineers, and which had a large overseas membership in the metal mining fields of South Africa and Australia. By the time that this national organisation was operating, tin and copper mining was already declining in the south western counties, and it is possible that the promotion of national activity weakened the provincial associations, but the region had supported a remarkable range of organisations in the nineteenth century, including the Royal Institution of Cornwall, established in 1818 as a 'literary and philosophical' society with the foundation of a Cornish museum as one of its objects; the

Royal Cornwall Polytechnic Society, founded in 1833; and the Royal Geological Society of Cornwall, which had its headquarters in Penzance. These were joined in October 1859 by the Miners' Association of Cornwall and Devonshire, the object of which was primarily educational - to raise subscriptions in order to promote classes in scientific subjects relating to mining. After two decades of modest success in this field,[33] the Association merged in 1885 with the Mining Institute of Cornwall, which had been established in 1876 with the more explicitly professional objectives of: 'considering, discussing and recording, mining matters and inventions, and suggestions for improvements in the practical and economical working and developing of mines generally'.[34] The joint body took the title of 'The Mining Association and Institute of Cornwall'.[35] It retained a strong interest in mining education, but it also gave regular attention to technical papers which were published in its *Transactions*, and negotiated to set up a Miners' Insurance Fund.[36] Despite these promising developments of professional activity, the new body did not survive long. The *Transactions* only ran to four volumes, the last being issued in 1895, and formal activities appear to have ceased. Subsequent attempts to revive professional organisation included a new society, the Cornish Institute of Mining, Mechanical and Metallurgical Engineers, which was formed in 1913 and continued to publish *Transactions* throughout the 1920s, but then seems to have faded away like its predecessors.[37] Cornish engineers, like other specialists and professional groups in the South West, did not lack in energy and enthusiasm, but with the continued decline in British metal mining they were short of numbers to sustain a permanent association of their own in the twentieth century, and it seems likely that most of those remaining sought refuge in the national organisation.

While the regional associations of professional engineers formed by mining engineers tended to adopt a missionary attitude to achieving improvements in safety and in standards of education, and survived into the twentieth century largely by virtue of their affiliations with national organisations, those promoted primarily by mechanical engineers tended to emphasise more modest aspirations such as the welfare of their members and convivial activities. They arose in areas such as the West Riding of Yorkshire, Manchester, and Birmingham, where there was a long tradition of machine making associated with textile machines, locomotives, and the manufacture of steam engines, and where there were many mechanical engineering companies ranging from the very large to very small establishments. Their appeal seems to have been primarily to foremen engineers and men of middle-management grades, with few of the senior engineering personnel becoming active members, which helps to explain the interest in benevolent funds for the benefit of members requiring assistance in adversity. Such associations flourished in the nineteenth century and acquired sufficient momentum to carry them on well into the twentieth century, but with falling recruitment and uncertain motivation they underwent a marked decline culminating in some cases in extinction.

The Leeds Association of Foremen Engineers and Draughtsmen was founded in 1865, drawing its membership from an impressive array of distinguished engineering firms in and around that city. Its title, together with its welfare provisions and its appeal to middle-management in its qualifications for membership - 'It is requested that each candidate for Ordinary Membership shall not exceed fifty years of age, and have been a principal foreman in some branch of the Engineering or Mechanical Trades for a period of two years'[38] - suggest that this was intended to be a sub-professional organisation. But the dividing line is not easily drawn, especially as the Association changed its title in 1890 to become 'The Leeds Association of Engineers' and modified its Rules at the same time so that: 'Employers and Managers may now be legally Ordinary Members'.[39] The Association also adopted some of the more distinctive policies of a professional institution by negotiating for some joint facilities with the Leeds University Engineering Society,[40] and by its decision to seek incorporation.[41] Moreover, the Association made a professional-style declaration of its objects in 1925, including the statement that: 'The Association does not discuss on any occasion the politics of the trade ...'.[42] At this time the total membership stood at 283, substantially less than the peak of 382 in 1921, but it managed to maintain an enrolment around this level until well after the Second World War. The Association seems never to have undertaken the publication of any *Transactions*, but it has entered into close relationships with other West Riding organisations of engineers, such as the Keighley Association of Engineers, founded in 1900, for joint meetings and convivial occasions.[43] The Leeds Association thus managed to make a modest contribution to the professional consciousness of engineers in the West Riding of Yorkshire.

Much the same can be said of the other institutions in this category. The Manchester Association of Engineers began in 1856 as 'The Manchester Association of Employers, Foremen and Draughtsmen of the Mechanical Trades of Great Britain', assuming its shorter title in 1885. In its early years, it met on Saturday evenings in the Merchants' Hotel, Oldham Street, growing from an attendance of about a dozen to a membership of sixty-seven in 1863 and of 178 in 1878.[44] Its first President was Charles Lister, a foreman from Galloways, and early members came from other prestigious engineering firms in the district such as Sharp Stewart and Beyer Peacock. In addition to regular meetings, when papers were read and discussed, the Association promoted visits, picnics, and an annual dinner, the latter normally being addressed by a distinguished local engineer.[45] The Manchester Association attached considerable importance to its Charitable Fund: the superannuation benefit was phased out in 1950, but the provision for immediate assistance to members in hard times remained.[46] The Association thus acted as a benefit society for its members in a much more direct sense than most professional institutions. On the other hand, the Manchester Association did have supporters amongst the professional engineers. One of these, Thomas Ashbury CE, was a consistent spokesman for the Association and twice became President: it was during his second term, in 1884-5, that the Association adopted its more professional title. Addressing the members in that year Ashbury observed optimistically:

> 'We meet under most encouraging circumstances. We have had during the past year forty-two new members, our finances are flourishing, our library increasing; we have removed to a most suitable room in a very appropriate building, the Technical School; and the retrospect of the past year and our present position, augurs well for an interesting and successful year.'[47]

The Association entered the twentieth century in a healthy condition, but then suffered the prolonged decline which afflicted most of the regional engineering associations. The membership at the end of 1970 was 375, and it continued to fall steadily to 234 in 1979,[48] at which time the Association was running a large deficit of £1,300 on the year's working, with printing being the largest single item. Like comparable societies in other regions, the Association was gradually fading away.

As in Manchester, so in Birmingham. Although it was the centre of Britain's largest manufacturing district outside the metropolis, with a strong engineering tradition, Birmingham did not have its own professional organisation of engineers until 1889. The presence of the Institution of Mechanical Engineers, with its headquarters in the city from 1847 to 1877, may have deterred local engineers from forming a regional association, but that body always aspired to be a national institution and made this clear by moving its offices to London.[49] When it began in 1889, the organisation styling itself 'The Birmingham and District Foremen and Draughtsmen's Mechanical Association' seemed to pitch its membership aspirations at the 'foremen of the metal trades' and other sub-professional echelons of engineering management.[50] However, a shift of emphasis towards a more professional outlook occurred quite early, being demonstrated in the change of title to 'The Birmingham Association of Mechanical Engineers' in 1891. It acquired the typical hierarchy of grades of membership of an engineering professional institution and set out its aims in characteristic professional terms:

'The Association is established to promote acquaintance and exchange of opinion between those engaged in, or connected with, the various branches of Engineering, and to assist in the promotion of Engineering Science in the City of Birmingham and the surrounding district. Also to form a fund for the benefit of its members under certain conditions as may be from time to time determined ... The Association does not discuss on any occasion the politics of the trade. Its objects are purely and solely philanthropic and instructive, and it essays to assist its Members in efficiently filling the important posts occupied by them.'[51]

In this form the Birmingham Association enjoyed an active existence until the Second World War, when it seems to have faded away. A series of published *Proceedings* ran from 1900 to 1940, and several papers which had been presented to the Association were published and bound together. Prizes were distributed to pupils at the Municipal Technical Schools,[52] annual dinners were addressed by distinguished persons, rooms and a library were established in the Grand Hotel, and the benevolent fund function does not seem to have received much emphasis.[53] Total membership reached 370 in 1905 and remained about that level until the 1920s, but declined thereafter. The centripetal tendencies of professional organisation were probably felt earlier in Birmingham, with its easy access to London, than in other parts of the country. They certainly proved to be overwhelming for the engineering organisation, the Birmingham Association slipping into an irreversible decline in the decades between the World Wars.[54]

Glasgow was the base for a sub-professional organisation of engineers similar to those in Leeds, Manchester and Birmingham. The Glasgow and West of Scotland Association of Foremen Engineers and Draughtsmen was formed in 1898:

'for the purposes of bringing together those engaged in the direction and superintendence of engineering works, for better acquaintance and promoting exchange of opinions on interesting questions constantly arising from the progress of mechanical trades.'[55]

The Association appears to have functioned as an engineering middle-management body. In the years between the World Wars it addressed itself to the problems of improving industrial relations.[56] It attained a membership of 586 in 1932. In 1939, with a slight decline of membership to 564, it adopted a simpler style as 'The West of Scotland Association of Foremen Engineers'. It is not clear from the available documentary evidence whether or not the Association survived the Second World War, but its public activities disappeared.

The Sheffield Society of Engineers and Metallurgists falls into the category of associations with a predominantly mechanical engineering character, although the emphasis in the city on the manufacture of cutlery and fine steel alloys contributed a strong chemical engineering or metallurgical component to the constituency of the membership. It was formed in 1894 by the merger of two previously existing organisations - the Sheffield Society of Engineers, founded in 1888, and the Sheffield Metallurgical Society, founded in 1890: 'as the objects of these societies were very similar, and a large proportion of the membership belonged to both ...'.[57] It does not appear to have produced the regular *Transactions* of a learned society, which was one of the common features of a professional institution, but several copies of its *Proceedings* have survived and individual papers were published occasionally, such as those by John Oliver Arnold, Professor of Metallurgy at the University of Sheffield, on the qualities of steel.[58] The Society was still in existence in 1949, when it was responsible for the installation of a plaque commemorating the birthplace on Attercliffe Common of Sir Robert A. Hadfield Bt. FRS.[59] No figures for the membership of the Society have been discovered, except for that of about 450 listed in 1914, and the later years of its history are particularly obscure.

It is unnecessary to be dogmatic about relative degrees of professionalism amongst the provincial organisations of nineteenth century engineers, but it seems likely that the bodies founded predominantly by mechanical engineers were closer to the tradition of Mechanics' Institutes than to professional institutions. Although they assumed some of the outward forms of professional societies, they catered for a somewhat different clientele with different objectives from those of aspiring professional men: they were, rather, foremen craftsmen concerned with extending their knowledge of their trade but without any expectation of elevation to senior managerial positions. They were more anxious about welfare benefits and pensions for their members than about innovative work on the frontiers of their discipline. It is a significant indication of this emphasis that none of them showed much interest in the publication of papers on original research, nor on the achievement of corporate status, at least until after the First World War. It is of interest that the national Institution of Mechanical Engineers only secured its charter in 1929.[60] Nevertheless, there are sufficient indications to show that the regional organisations of mechanical engineers made a contribution to the prestige and self-esteem of the emerging concept of middle-management in the British engineering industry, even though it is frustratingly impossible to quantify this contribution in a precise manner. They deserve to be recognised, therefore, as part of the network of engineering professional organisations.

The most professionally conscious of all the regional organisations amongst British engineers were those initiated by civil engineers and shipbuilders. Four of these deserve consideration: those established in Glasgow, Manchester, Newcastle and Liverpool. In the middle of the nineteenth century Glasgow was well placed to become the base for a strong regional association of professional engineers. Nevertheless, it required an external catalyst to inspire the formation, in 1857, of the Institution of Engineers in Scotland. This was acknowledged in the preface to the first volume of the *Transactions* of the Institution, where it was observed that:

'it owed its origin, in a great measure, to the very successful Meeting held in Glasgow, in 1856, by the 'Institution of Mechanical Engineers' of Birmingham: the great utility and probable success of a similar Institution holding its meetings in Glasgow ... being at the time strongly impressed upon several Engineers residing in and near Glasgow... [61]

Amongst the Glasgow engineers, one of the strongest advocates for a regional institution was W. J. Macquourn Rankine, who had been appointed Professor of Civil Engineering at Glasgow University in 1855 and who was elected the first President of the Institution in 1857. Rankine presented a carefully formulated 'Introductory Address' in which he stated explicitly the intention of the new organisation to be inclusive rather than exclusive as far as different sorts of engineering specialisation were concerned. He concluded by stressing the advantages of Glasgow, describing it as the 'metropolis of engineers': 'If an institution of engineers is to make good progress anywhere', he argued, 'it ought to be in Glasgow'.[62]

The Glasgow Institution certainly got off to a good start. Although the meeting of the Institution of Mechanical Engineers in the city had provided the initial inspiration for organisation, the new association recruited its members from all types of Scottish engineers. By the end of its first session, in April 1858, it already had a membership of 127, and a year later this figure had risen to 164, which included eight Honorary Members of remarkable catholicity: the Earl of Dundonald, James Walker, James Prescott Joule, C. P. Smith, William Fairbairn, William Thomson, Robert Stephenson, and Professor R. Clausius.[63] The total membership increased steadily thereafter until the First World War, although there were some unusually high accessions in the years when other associations merged with the Institution. The most substantial of these amalgamations was in 1866, when the Scottish Shipbuilders' Association joined the Institution. This had been established in 1860, presumably because a significant number of shipbuilding engineers felt that their specialist interest was not being adequately represented by the Institution, even though it had been specifically designed to include them. It may be also that the Glasgow shipbuilders were inspired to emulate the Institution of Naval Architects set up in the same year in London. Whatever the reasons for its foundation, the Scottish Shipbuilders' Association enjoyed six years of independent life, listing 144 members under the presidency of Robert Barclay in 1862, and publishing weighty *Transactions*, the first volume of which contained the seminal treatise by David Kirkaldy on 'Results of an experimental enquiry into the comparative tensile strength and other properties of various kinds of wrought-iron and steel'.[64] With the amalgamation in 1866, the joint body applied for corporate status under the title 'The Institution of Engineers and Shipbuilders in Scotland'.[65]

Incorporation was achieved in 1871, with a membership of about 400. By 1900, under the presidency of Robert Caird, it had reached 1,243, and at the outbreak on war in 1914 it stood at 1,677. The Institution then occupied luxurious premises in Elmbank Crescent, where a memorial tablet was erected in 1913 to commemorate the engineers who had died in the S.S. *Titanic*.[66] Membership reached an all-time peak of 1,847 in 1920. Thereafter a steady decline set in through 1,675 in 1925 and 1,166 in 1935 to the Second World War, when the fall was temporarily arrested, only to resume in the 1950s. The figure had fallen to 1,063 in 1972, the last figure in the series of *Transactions* in Glasgow University Library. The Institution adjusted to harder times by disposing of its Elmbank Crescent property in favour of a more modest office in Bath Street, which was renamed 'Rankine House', and by presenting its substantial library to the University of Glasgow. Although it still possesses a considerable number of members, it would seem fair to say that, like similar bodies in other parts of the country, the Institution had completed the most professionally useful part of its life by 1970.

In its prime, the Institution of Engineers and Shipbuilders in Scotland was one of the largest and most influential of the regional organisations of British engineers. The list of its Presidents is an impressive array of British engineering talent, including W. M. Neilson, J. P. Napier, W. Denny, A. C. Kirk, and Sir William Arrol, by the end of the nineteenth century. The Institution even acquired international prestige, by taking advantage of the Engineering Exhibition in Glasgow in 1901 to mount an International Engineering Congress. This was held from 2nd to 5th September 1901, and it seems to have been a genuinely pioneering effort. To judge from the list of societies represented, foreign and colonial delegations, and individual membership, the Congress achieved a large attendance drawn from all parts of the world. The Institution of Civil Engineers gave its general blessing to the Congress, but the Glasgow Institution was responsible for the organisation, and the Congress provided an unprecedented opportunity for the world's engineers to compare their experiences.[67]

After this success story, the history of the Manchester Institution of Engineers is a remarkable disappointment. Even allowing for the success of the Manchester Association of Engineers in providing for the middle ranks of engineering management, there ought to have been sufficient resources amongst the thriving engineering industries of the region to support a more overtly professional organisation, and the foundation of the Manchester Institution of Engineers in 1867 appeared to signal a recognition of this need, but the appearance belied the reality, and there is little to be said about the Institution. The available details of the organisation are frustratingly incomplete, with only two volumes of *Proceedings* covering the period from November 1867 to December 1868 surviving in the Manchester Library Collection, so that it would seem that the Institution did not have a long life. Amongst the policies of which records exist, however, were such characteristic expressions of professional consciousness as support for the local university, marked in this case by the eagerness with which the Institution supported 'a professorship of engineers' at Owens College:

'It is thought desirable and practicable to establish in some way a connection between this institution and the college by which the objects of both may be advanced.'[68]

Engineering studies became strongly established in Manchester under Professor Osborne Reynolds, but support to sustain the Institution of Engineers appears to have disappeared.[69]

The fate of the North East Coast Institution of Engineers and Shipbuilders was different again. Professional civil engineers and shipbuilders on Tyneside did not fit comfortably within the predominantly coal-mining North of England Institute of Mining and Mechanical Engineers, and decided to take action on their own account. The initiative came in 1884 from a group of young engineering draughtsmen in the shipbuilding industry, but when one of these, W.G. Spence, canvassed for support in the columns of the *Newcastle Daily Journal*,[70] he received strong encouragement from William Boyd, manager and director of the Wallsend Slipway and Engineering Co. Ltd., and other senior engineers. The result was the formation, on 28th November 1884, of the Institution, with Boyd as its first President. The model of the new organisation seems to have been the Institution of Engineers and Shipbuilders in Scotland,[71] although there was also some recognition of the Institution of Naval Architects as 'the parent institution'.[72] Certainly its objects - 'the advancement of the sciences of engineering and shipbuilding' - reflected those of existing professional institutions, and in fulfilling them the North East Coast Institution has enjoyed considerable success. Its *Transactions*, from 1884, have maintained a high standard of technical excellence and have given prominence to many important innovations in marine engineering, including the triple expansion steam engine and the steam turbine, the latter having been developed by Sir Charles Parsons who delivered what is believed to have been his first published paper to the Institution in 1887.[73]

The Institution built up a large library, recently donated to Sunderland Polytechnic, an active Graduate (later, 'Student') Section, and a record of vigorous activity in technical education through scholarships and the promotion of a chair of engineering and naval architecture at the College of Physical Science in Newcastle.[74] It has co-operated extensively with other institutions, and has supported the project for a scientific and industrial museum in Newcastle. It was incorporated in 1914 and received Royal Patronage at the same time. The first membership list of the Institution recorded 373 Members, thirty-nine Associates, and forty Graduates. Numbers rose fairly gradually to a peak of 2,367 members in 1960, but in recent years they have fallen back to about 1,400.[75] In this respect it has suffered from the centripetal tendencies of professional organisations in the twentieth century, as well as the change in the industrial character of North East England, with shipbuilding much less prominent than it was fifty years ago. As an affiliate member of the Council of Engineering Institutions (CEI), the membership qualifications of the Institution have been brought into line with those of the national body. In one way the Institution scored a notable 'first' amongst engineering professional organisations - it specifically admitted women to its membership in 1919.[76]

In Glasgow and in Newcastle, the success of regional professional associations of engineers was derived largely from the support of members employed in shipbuilding. In Liverpool, the last case we have to consider, it came primarily from civil engineers engaged in public works - particularly docks and water supply. The Liverpool Engineering Society was formed in November 1875. Like the North East Coast Institution a decade later, and like the venerable precedent of the Institution of Civil Engineers in 1818, the initiative came from some junior engineers: in this case, from six engineering pupils of G. F. Lyster, who was then the Engineer-in-Chief to the Mersey Docks and Harbour Board. The young men styled themselves 'The Liverpool Engineering Students Society' and met weekly to present papers to each other. In response to interest from older practitioners, they readily tempered their enthusiasm with experience by extending their membership to include senior engineers, and by omitting the word 'Students' from their title.

By a happy accident, a good proportion of the historical records of the Liverpool Engineering Society have survived in Liverpool City Archives.[77] Together with the impressive series of eighty volumes of *Transactions*, covering the years from 1881 to 1960, these make possible an unusually comprehensive insight into the rise and decline of a provincial engineering professional institution. Half a century after the foundation of the Society, W. E. Mills presented a paper in its *Transactions* reviewing its history. Mills had missed being a founder-member by four years, but with the experience of forty-six years of membership behind him, including a year as President in 1885-6, he was able to give a well-balanced survey with some interesting anecdotes. He recalled congratulating the Society 'and the Engineering Body in Liverpool in general' on the establishment of a chair of engineering at Liverpool University in the year of his presidency. The first occupant of the chair, Professor H. S. Hele-Shaw, went on to become President of the Society in 1894. Mills reviewed an impressive selection of the papers which had appeared in the *Transactions*, and the contribution made by members to a range of national panels and study groups.[78]

The Society had good reason to be satisfied with its growth in the first half-century of its existence. In 1880 it already had seventy-six members (three Honorary and seventy-three Ordinary), and the figure continued rising to 110 in 1885, 483 in 1901, and to an all-time peak of 886 members in 1925 with only a temporary check in the First World War. Thereafter, however, there was a prolonged decline in membership, only briefly arrested in the years immediately after the Second World War, to 510 in 1959, representing the lowest figure since 1902.[79]

By 1960, the critical mass of the membership was no longer able to sustain the facilities which the Society had provided for many years. The contraction in numbers thus brought a rapid falling off in services. First to go was the publication of the annual *Transactions*. This distinguished work was replaced by the *Journal*, a flimsy magazine devoted mainly to engineering advertisements although it also served as a medium for the Annual Report and other official notices. The Annual Report for 1962-3 showed that the Society, with a deficit of £489 on the income and expenditure account, had still not balanced the losses incurred largely through its previous publications, and the membership had sunk to 472. Further economies were thus required. The Society, which had operated for many years as a 'gentlemen's club' for Liverpool engineers, through rooms in the city providing a service of meals and conviviality, decided to abandon both the meals service and the rooms. Then in 1966 the appearance of a branch of the CEI on Merseyside, offering alternative facilities for professional engineers, was greeted with anxiety: one of the last Presidents of the Society, Captain D. A. Smith, spoke of the CEI as 'the villain of our piece'[80] and concluded gloomily:

'Since I came to Liverpool in 1946, I have seen the inexorable growth of the influences which have undermined the position of the Society ...'[81]

A year later, Captain Smith's successor gave what was to be the last New Year Message
to the Society. He declared himself to be happy about the prospects for the immediate
future:

'We had a most enjoyable Theatre evening in October ... This was followed by a ladies' evening in November ... Dancing, Bingo, and a happy time for everyone ...'[82]

This is a sad epitaph for a Society which, after almost a century of serving the interests of
engineers on Merseyside so well, died with a whimper rather than a bang, by ceasing to
function in 1968. But it provides a particularly vivid illustration of institutional redun-
dancy in the engineering profession, and it can be taken, with caution, as representing in
microcosm all the general influences affecting the regional diversification of the profession.

The regional engineering associations which have been reviewed here have varied
considerably in their aims, their organisation, and their policies. These variations have
been conditioned in part by regional differences between the engineering industries, although the precise relationship between these industries and the professional associations of each area remains largely one of tantalising obscurity because of the lack of
adequate prosopographical evidence. What is clear, however, is that the pattern of institutional development has been deeply influenced by the type of engineering dominant
in each region with marked differences of outlook according to whether promoted by
mining engineers, by mechanical engineers, or by civil engineers and shipbuilders. Also,
a general pattern of development has emerged over and above these differences: all the
regional institutions were established and flourished in the second half of the nineteenth
century: all declined in the first half of the twentieth century: and few of them survived
into the second half of the twentieth century, and then only in much reduced circumstances. We can be reasonably sure that the other regional organisations which, for lack of
information, have been excluded from this survey, do not depart to any significant extent
from this pattern, and we can be similarly confident that the survey has covered, however
superficially, all the major engineering districts of Britain. There is, therefore, a real problem requiring historical explanation about this pattern of rapid institutional growth in the
regions followed by a spectacular collapse. Why did it happen?

The phase of growth is comparatively easy to explain, as it can be safely assumed that
the same economic and social forces which were conducive to the proliferation of national professional institutions of engineers in the second half of the nineteenth century
were applicable also in the regions. That is to say, the impact of improved transport and
communications, the general rise in the prosperity of the nation and the standard of living of all social classes, and the expansion of engineering into new areas of professional
competence, did as much to encourage engineers to organise themselves for instructional
and social purposes in the regions as they did at national level. The only significant differences were, first, in the scale, with regional members being numbered in hundreds
rather than the thousands who frequently swelled the ranks of the national institutions;

and second, and partly in consequence of this difference in scale, the national institutions showed more scope for specialisation than was possible at a regional level. Whereas the process of national institutional proliferation was towards increasingly refined degrees of specialisation, the local organisations depended on the support of all types of engineers within their territorial catchment areas.

The more difficult historical problem is that of accounting for the rapid decline of the regional associations in the twentieth century, at a time when the national organisations of the engineering professions have continued to grow with remarkable vigour. At one level, this can be explained in terms of the loss of many of their distinctive functions as a result of more general economic and social change. The aim of improved ventilation in coal mines, which had been such an important objective of the nineteenth century coal mining engineers, has been fully achieved. Similarly, benevolent club functions have been made largely redundant by the extension of the social services, and the urge to widen educational horizons has been fulfilled in new university engineering schools and other provision for technical training. But these changes affected national as well as regional associations, so it is necessary to look elsewhere for a full explanation of the differential development between them. In particular, the structural changes implied in the use of the image of 'centripetal tendencies' must be explored, as it is these which have been responsible for a shift in the balance between national and regional societies. In the first place, there was a distinct change in emphasis as the leading national engineering institutions recognised that the creation of increasingly specialised organisations could be prevented by the creation of special departments or 'sections' within the parent institutions, thus preserving their own numbers and avoiding unnecessary competition. The Institution of Mechanical Engineers, for instance, after watching the establishment of institutions for Automobile and Locomotive engineers, took steps to accommodate these and brought about successful amalgamations.[83] This policy affected regional organisations when the national bodies came to understand that they could further strengthen their recruitment and serve their members by the establishment of provincial branches, for when these branches came into competition with existing regional institutions it was the branches, possessing the resources, prestige, and personal contacts of national organisations, which had all the advantages and achieved the dominant position.

Another aspect of this centripetal tendency was the increasing demand for specialist qualifications. In the nineteenth century, British engineers and their institutions retained their traditional insistence on practical experience as the only valid training for entry to the profession. But gradually the emergence of scientific engineering, requiring theoretical skills inculcated by university type instruction, put a premium on qualifications - university degrees, national diplomas, City and Guilds examination certificates - which were nationally recognised, and when the national organisations began to incorporate such qualifications in their own requirements for membership, and to devise their own examinations to test the competence in these fields of candidates for admission, the regional organisations found themselves at a new disadvantage. Faced with the choice between a regional society with no stringent examination requirements, and a national institution stipulating such requirements which would then be recognised throughout the profession, most able young engineers chose to give their support to the latter. Of course, many could and did support both, but the regional societies could not hope to survive merely as convivial organisations at a time when the national institutions were providing full professional qualifications and a viable forum for instruction and personal contacts. The regional societies simply could not match this competition except at the most trivial level of social intercourse, and at this level modern industrial life offers many alternative diversions and organisations, so that the societies lost out completely.

Yet a further aspect of the centripetal tendency which adversely influenced the position of the regional institutions was the diminution of their role as instructional organisations. The performance of associations of engineers as sources of mutual instruction through the presentation, discussion, and publication of learned papers, has always been one of their most distinctively professional functions, and many regional associations maintained a high level of achievement in this respect, at least well into the twentieth century. However, the increasing specialisation of engineering made it more and more difficult to keep up a high level of technical presentation in societies which, by their nature, were territorial in their recruitment rather than specialist, so that the high degree of technical sophistication required for the useful discussion of a specialist paper became unobtainable in the regional societies, which preferred a more generalist and therefore a less innovative approach. In other words, the frontiers of engineering knowledge and experience have been explored almost exclusively through the specialist societies and their various sub-sections, while a more bland presentation has become appropriate to the regional societies. Consequently, the regional societies have become diminishing attractions in the twentieth century as far as enterprising young engineers have been concerned, because such men have rightly seen their professional careers being enhanced by their contributions to the highly technical discussions of the national specialist societies. Membership of the regional societies became for these men an optional extra, which might offer some social advantages but which had little of professional value for the ambitious engineer.

These centripetal tendencies were consolidated by the increasing co-ordination between the national institutions. It is significant in this respect that the beleaguered leadership of the Liverpool Engineering Society, in the final stages of its existence, saw the CEI as a particular cause of its difficulties. The CEI represented a step by the national bodies to transcend their specialist differences, so that when it began to organise regional branches it did so on the same sort of cross-disciplinary basis as that which had previously been one of the great strengths of the regional societies. Whereas branches of the Mechanicals or Structurals preserved a specialist appeal, branches of the CEI attracted engineers of different specialisations and thus the regional societies lost their last claim to distinctive professional existence.

The disappearance of the regional organisations of engineers may be regretted, along with other indications of local enterprise and grass-roots vitality, but it is futile to deplore it. The greater efficiency offered by the national organisations, once they had recognised the importance of educational qualifications and the merits of a flexible sectional and branch system, so outweighed the administrative powers of regional societies that the move towards national co-ordination and control became inevitable. What remains to be seen is whether or not the logic of this development - towards one large organisation replacing specialised institutions as well as regional societies - will be pursued by the British engineering profession. The signs at present suggest that this will be a long and slow process.

Notes

1. No good history of the Institution of Civil Engineers of Ireland has been written. But see Noel J. Hughes, *Irish Engineering 1760-1960*, Institution of Engineers in Ireland, 1982 - mainly an extended bibliography - and Ronald C. Cox (compiler), *Engineering Ireland 1778-1878*, Exhibition Catalogue, Trinity College Dublin, 1978.

2. Hughes, *op. cit.* 59-65.

3. Cox, *op. cit.* p.6.

4. N. O'Dwyer, Presidential Address to celebrate the Centenary of the Institution in 1935. *Trans. Inst. Civil Engs. of Ireland*, 62, 1936, 3.

5. Hughes, *op. cit.* 69-71: see also G. W. Hemans, Presidential Address, *Trans. Inst. Civil Engs. of Ireland*, 5, 1860, 51-65, esp. 55.

6. R. Mallet, Presidential Address, *Trans. Inst. Civil Engs. of Ireland*, 8, 1868, 48-102, esp.49.

7. *Trans. Inst. Civil Engs. of Ireland*, 1, 1845: the first paper was by R. Mallet, 'On the Artificial Preparation of Turf, independently of Season or Weather' - 1-49, i.e. a method of drying peat in a kiln.

8. Cox, *op. cit.*, gives a list of the presidents up to the charter of incorporation in 1877, and biographical details on most of the fourteen named.

9. Mallet, *op. cit.* p.52.

10. Ibid., p.90.

11. Michael J. Buckley, Presidential Address, *Trans. Inst. Civil Engs. of Ireland*, 56, 1931, 3, gives a review of membership figures from which he concludes: 'that there still remain a regrettably large number of engineers who are not alive to the benefits of membership of our institution'.

12. Another Irish regional institution was established in Belfast in 1892: see D. B. Marr, *Recollections and Memoirs of the Belfast Association of Engineers*, Belfast 1967 - a very slight twenty-five page pamphlet. No figures of membership or other vital statistics are given.

13. Other regional engineering institutions include the Bolton Engineering Association, the Hull and District Institution of Engineers and Naval Architects, and the Cleveland Institution of Engineers, but I have not been able to establish any significant details about these. Various London-based regional organisations also remain shadowy presences.

14. Quoted in the *Centenary Brochure 1852-1952*, The North of England Institute of Mining and Mechanical Engineers, Neville Hall, Newcastle-upon-Tyne, 16 pages, p.7

15. *Council Minute Book*, North of England Institute of Mining and Mechanical Engineers, in the Archives of the Institute, 1, 29 January 1853 and 4 January 1855.

16. Ibid., 1, 28 March 1857, 7 May 1857, etc.

17. *Trans. North of England Institute of Mining and Mechanical Engineers*, 1, 1852-53, 3-23. For discussion of exchange of information with the Mining Association of Hainault and with other bodies in Belgium and France, see *Council Minute Book* 1, 30 January 1864, 3 September 1864, etc.

18. The Institution of Mining Engineers was formed by four local associations in 1889. Until 1907, when it moved to London, the head office remained at Neville Hall in Newcastle. See above, Chapter 5.

19. See: 'Notice of the Preliminary Proceedings which led to the formation of the Federated Institution of Mining Engineers' in *Trans. Inst. Min. Eng.*, 1, 1889-90, viii-x.

20. Verbal communication with the Secretary of the Institute for the 1981 figure.

21. *Centenary Brochure 1857-1957*, South Wales Institute of Engineers, Cardiff, 1957, 58 pages, p.6.

22. Ibid.: the last three words in parentheses were omitted from the revised form of this statement.

23. Ibid., p.11.

24. Ibid., p.15: the library block was added in 1907.

25. Ibid., p.30.

26. A mining department was started at the University College of South Wales (Cardiff) in 1891, and local coal owners set up a rival college at Treforest in 1912.

27. F. B. Hinsley: 'A Centenary History of the Chesterfield and Derbyshire Institute of Mining, Civil and Mechanical Engineers and its Successors', in *The Mining Engineer*, 1976, pt. I (1871-1889), May 1976, 493-500; pt. II (1889-1946), June 1976, 561-564; and pt. III (1946-1971), July 1976, 623-627.

28. Ibid., 493-4.

29. Ibid., p.496: the controversy was known as 'The Battle of the Fans'.

30. Ibid., p.500.

31. Ibid., p.625.

32. The Institution of Mining and Metallurgy was formed at a 'Preliminary Meeting' held in London on 13 January 1892, at which Mr. Charles Seymour, CE, presiding, observed that no existing institution, with the honourable exception of the Mining Association and Institute of Cornwall, had hitherto devoted itself to non-ferrous metal mining and processing - see *Trans. IMM*, 1, 1892-93, 34. See above, Chapter 6.

33. See the 'Reports' on the Annual Meeting, *Proceedings*, the Miners' Association of Cornwall and Devonshire, I, 1866-1873 and II, 1874-83.

34. *Proceedings, Mining Institute of Cornwall*, 1, December 1876-December 1883, 222

35. *Trans. Mining Association and Institute of Cornwall*, 1, 1885-7

36. Ibid., 1-3, especially 3, 1894, p.172.

37. *Trans. Cornish Institute of Mining and Mechanical Engineers*, 1, 1913 et seq.

38. Leeds Association of Engineers, *19th Annual Report*, 1884.

39. Ibid., *25th Annual Report*, 1890.

40. Ibid., *49th Annual Report*, 1914; this agreement was, however, abandoned on account of the outbreak of the First World War.

41. Ibid., 1923 (first in new format), p.18.

42. Ibid., 1925: 'Objects of the Association'.

43. After dealing with a number of engineers' institutions which had become moribund, it was refreshing to find a clutch of active associations in the West Riding, issuing programmes of attractive meetings, and I am particularly grateful to Mr. J. A. King, Hon. Secretary of the Leeds Association of Engineers, and to Mr. H. Forsyth, President of the Keighley Association of Engineers, for kindly answering my enquiries about their organisations.

44. *The Manchester Association of Engineers 1856-1956: a hundred years of Engineering in Manchester*, Manchester, 1956.

45. Guest speakers in the years before 1885 included Sir William Fairbairn, Richard Peacock, William Mather, and Daniel Adamson, all of them prominent entrepreneurial engineers in the Manchester district. See A. C. Dean, *Some episodes in the Manchester Association of Engineers*, Manchester, December 1938, p.57, for an account of Adamson's address in 1881.

46. The role of the Charitable Fund is discussed in the previous two references: *A hundred years* ... and A. C. Dean, *Some episodes* ...

47. *Address of Mr Thomas Ashbury, CE, on his being elected for the second time President of the Manchester Association of Employers, Foremen, and Draughtsmen of the mechanical trades of Great Britain*, Manchester, 1884, 35

48. *Trans. Manchester Association of Engineers*, 1978-79, 123rd Annual Report: the number actually marked a slight rise of two in the previous year, but the overall trend was unmistakable.

49. R. H. Parsons, *History of the Institution of Mechanical Engineers, 1847-1947*, published by the Institution, 1947, 28-9.

50. *Proceedings, Birmingham Association of Mechanical Engineers*, for Session 1928-29: see the historical review in the Presidential Address by Donald G. Mackintoch, MBE, MIMechE, MIStructE.

51. Ibid., *Proceedings*, 1905 - the last item (644366) in the bound 1900-1905, in Birmingham City Library, L65.206

52. They amounted to £5.5.0d, in 1905.

53. Grants totalling only £50 were distributed to members and their dependents in difficulties in 1930-31: *Proceedings*, Report on Session, 99-104.

54. A card from the Secretary, Walter Hadley MIMechE, to the City Librarian, dated 20 January 1942, is bound in the last of *Proceedings* for 1939-40 in Birmingham City Library: it promised to send future issues as they were published, but no more appeared.

55. Objects of the Association, stated in the *Annual Reports* of the Glasgow and West of Scotland Association of Foremen Engineers and Draughtsmen, for which a few s for 1931-32, 1938-39, 1940-41, and 1943-44, exist in the Mitchell Library, Glasgow.

56. See pamphlets on 'Goodwill in Industry', 1928, and 'Paths to Peace in Industry', 1930.

57. Sheffield City Local History Library: *Local Pamphlets* 156, no. 2: 'Rules and List of members' of the Sheffield Society of Engineers and Metallurgists, 1914, p.iv.

58. Ibid., *Local Pamphlets*, 179, no. 12; and 153, no. 8.

59. Ibid., *Local Pamphlets*, 179. no. 10.

60. R. H. Parsons, *A History of the Institution of Mechanical Engineers 1847-1947* published by the Institution, 1947, p.66

61. *Trans. of the Institution of Engineers in Scotland*, 1, 1857-58, preface, August 1858.

62. *Trans. Inst. Engs. in Scotland*, 1, 1857-58, 3. It appears that Rankine had had a disagreement with the Institution of Civil engineers over its failure to transfer him to the grade of Member despite his occupation of the Chair at Glasgow, the Annual Report of the Civils for 1857-58 noting his resignation from the Institution as an Associate - see Hugh B. Sutherland, *Rankine - His Life and Times*, the Rankine Centenary Lecture at the University of Glasgow, 1972, p.11.

63. *Trans. Inst. Engs, in Scotland*, 1, second session 1858-59.

64. *Proceedings of the Scottish Shipbuilders' Association*: this massive Report by Kirkaldy, consisting of 212 pages including many tables and appendices with sixteen plates, is sandwiched between the First Session (1860-61) and the Second Session (1861-62). The experiments were originally promoted by Robert Napier, in whose employment Kirkaldy had spent nineteen years, and the first part had already been published in *Trans. of the Inst. of Engs. in Scotland*, 2, 1859.

65. This title had been the subject of some argument, but by 1870 it seems to have been accepted happily by all parties.

66. *Trans. Inst. Engs. in Scotland*, 57, 1913-14, 671.

67. *Report of the Proceedings and Abstracts of the Papers Read - International Engineering Congress (Glasgow) 1901*, ed. J. D. Cormack, Glasgow, 1902, 406 pages

68. *Proc. Manchester Inst. of Engs.*, Pt. II, President's Address by W.W. Hulse.

69. For Reynolds' career at Owens College, see Robert H. Kargon, *Science in Victorian Manchester: Enterprise and Expertise*, Manchester, 1977, pp.182-190: also Chapter 9, below.

70. *Newcastle Daily Journal*, 17 March 1884: the letter, signed 'T. square', is reprinted together with the reply from W. Boyd, at the end of Boyd's inaugural address as president of the North-East Coast Institution of Engineers and Shipbuilders, *Transactions*, 1, 1884-85, 3-19, 15.

71. See Spence's letter, and T. S. Nicol, 'The Institution', *Trans. N.E. Coast Inst. Engs.*, 77, 1961, 151-166, 152.

72. W. H. White commenting on Boyd's inaugural address, *op. cit.* 1884-85, p.17.

73. T. S. Nicol, *op. cit.*, p. 160.

74. Ibid., 156-7: the first appointment to the chair was Professor R. L. Weighton, who was nominated by the Institution, which guaranteed a substantial proportion of his salary, in 1886: he held the post until 1920.

75. Figures from *Trans. N.E. Coast Inst. Engs.*: verbal communication with the Secretary for the more recent figures.

76. T. S. Nicol, *op. cit.*, p. 156. But the Royal Aeronautical Society had admitted women in 1898.

77. Dr. S. J. Kennett, Senior Vice-President of the Liverpool Engineering Society in its final operative session, 1967-68, deposited the records of the Society in the Liverpool City Archives and Local History Library in February, 1973. They amount to some thirty s and bundles, classified under 620 ENG-Acc. 2561.

78. W. E. Mills, 'Fifty Years of the Liverpool Engineering Society', in *Trans. Liverpool Engineering Society*, 47, 1926, 171-204.

79. Ibid., 80, 1960, 112-13, giving an excellent tabular summary of the membership position.

80. *Journal*, Liverpool Engineering Society, 12, Jan. 1967, no. 5.

81. Ibid.

82. Ibid., 94 (the series had been renumbered to include the series of *Transactions*), Jan. 1968, no. 3. The last President was Harry Rostrow.

83. The Institution of Automobile Engineers merged with the Mechanicals in 1946, and the Institution of Locomotive Engineers did likewise in 1969.

Chapter Eight

British Engineering Abroad

While the engineering profession grew rapidly in the nineteenth century, and institutions proliferated to cater for the national and regional requirements of its members, there was an increasing flow of British engineers going abroad in search of fame and fortune. There had been a time in the sixteenth and seventeenth centuries when Britain had been dependent upon engineering expertise acquired from other countries, as in the case of German miners operating in the Lake District and elsewhere, of Huguenot silk and paper manufacturers, and of Dutch drainage engineers like Cornelius Vermuyden. In the middle of the eighteenth century it was still a foreign expert, the Swiss engineer-architect Labeyle, who had been called in to build Westminster Bridge.[1] But thereafter, as we have seen, a competent body of native engineers emerged and there was less reliance on outside talent. British engineering was at first home-spun and self-reliant: not only did it not receive any significant input from outside, but it also made little contribution itself beyond the shores of Great Britain. British engineering was striking strong roots at home, but it took some time for it to prepare itself for an expansive movement overseas. This relative isolation accentuates, in retrospect, the contrast with the subsequent period, when British engineers were willing and able to take their skills to the four corners of the Earth.

The fact that the isolation of British engineering in the period before 1830 was only relative should serve to remind us that there were exceptions, and some of these were sufficiently important to be worth mentioning as indications of the directions which British engineering penetration would later take. In the first place, the international interest in British steam technology led to a small but significant movement of engineers and their machinery from quite early in the eighteenth century. The first Newcomen engine in Sweden was erected at Dannemora iron mine in 1727, and the Potter family was busy building atmospheric steam engines in various parts of Europe about the same time.[2] In 1753, Josiah Hornblower emigrated across the Atlantic and built the first steam engine in the American colonies, a few years before Samuel Slater brought his vital expertise in the new cotton spinning technology to New England.[3] In several parts of the world, therefore, British mechanical innovations were already arousing considerable attention in the eighteenth century, and encouraging thereby an outward flow of native talent.

Secondly, although the British canal builders from James Brindley through to the end of the canal boom in the second decade of the nineteenth century took little interest in practice overseas, there were some notable exceptions. John Smeaton made a brief excursion to the continent to view civil engineering works in Holland and Flanders in the

1750s, but undertook no foreign commissions.[4] More significantly, Thomas Telford was persuaded by Baron von Platen to build the Götha Canal across southern Sweden, for which he was honoured with a Swedish knighthood.[5] About the same time, Lt. Col. John By of the Royal Engineers supervised the construction of the Rideau Canal through very similar terrain (and for similar strategic reasons) across the Laurentian Shield between Ottawa and Kingston in Ontario.[6] Even earlier, William Weston had gone to America in 1792 to advise on canal construction.[7] So despite the insularity of most British canal engineering, there was some transfer of the new transport technology by British engineers before the advent of the railways.

In addition to these small but not unimportant instances of the transfer of British mechanical and civil engineering skills, there were a handful of cases of individual British engineers in the eighteenth century going to practise overseas. The example of Josiah Hornblower has already been mentioned: he had only gone out to America to advise on the construction of steam engines, but he had such a terrible voyage out that he decided to stay in America rather than risk the return trip, and lived on as a prominent citizen of New Jersey until his death in 1809.[8] Another British emigrant to the United States was of even greater moment. Benjamin Henry Latrobe came from German Moravian stock, but he was born and brought up near Leeds in Yorkshire. He probably worked under Smeaton, and certainly assisted Jessop on the Basingstoke Canal. Then in 1796 he went to the United States, becoming engineer for the Philadelphia water works and architect of the Capitol. He died at New Orleans in 1820. It has been claimed for Latrobe that he was the father of both engineering and architecture in the USA.[9]

The drift of individual engineers overseas continued into the beginning of the nineteenth century. Some of the more visionary British engineers in the 1820s recognised that the development of the superb Cornish engines for mine pumping provided an opportunity to re-open the legendary gold and silver mines of Central and Southern America which had been abandoned as unworkable. Companies were floated to promote such ventures, and Richard Trevithick spent some of the best years of his life in pursuit of this El Dorado, returning to Cornwall in 1827 a broken man.[10] The trouble was that, even though the technology was adequate to the task, the operation of shipping out large steam engines, reassembling them on site, and then working them with local labour, proved to be beyond the organisational capacity at the disposal of the engineers in the 1820s, even though it became available in time for the gold rushes in the middle of the century. Strangely, Robert Stephenson met Trevithick in South America when he spent three years there (1824-27) in pursuit of the same chimera. But Stephenson was then a young man, with the great achievements of his career before him, so that his lack of success in South America was a stimulus rather than a handicap to him.[11]

Other British engineers who established careers for themselves in foreign countries during the eighteenth and early nineteenth centuries included a number who helped to transfer the new textile, iron, and machine technologies to Europe, amongst whom the Cockerill Family was one of the more successful. William Cockerill went to Sweden with his sons William and James in 1797, but the family soon moved to Belgium, where they built up a flourishing textile machine manufacturing firm at Verviers. The sons later went on to develop an interest in iron founding and general machine tool work, which played an important part in the industrialisation of Belgium.[12]

Although all these exceptions are of interest, the generalisation that British engineering before 1830 was somewhat parochial remains sound. It would be difficult to discern any over-riding trend from these few instances of British engineers seeking their fortunes abroad: there was certainly as yet no out-flow of engineering talent. It was the boom in railway construction which brought the first serious break in the insularity of British engineering and precipitated the swarm of engineers from Britain to foreign and colonial adventures. The first complete railway, offering time-tabled transport for goods and passengers, was the Liverpool and Manchester Railway, opened to traffic in 1830. The success of this line, operated by the 'Rocket' style locomotive built by the Stephensons, prepared the way for the tremendous wave of organisation, capitalisation, and railway construction which followed in the next three decades, and which left Britain and many other parts of the world with a well-developed railway system. Professional engineers played a vital part in this programme of railway building, both in the organisation, which included promoting acts of parliament before parliamentary committees, and in supervising the constructional works, as well as in the supply of locomotives, rolling stock, and operating equipment. The demand for more and more qualified engineers placed a heavy burden on the Institution of Civil Engineers, which had been founded in 1818 and which, under the benign presidency of Thomas Telford, had grown soundly and substantially. But Telford lived too early to be a railway man, and in any case he died in 1834, without making any contribution to the Railway Age. The Institution was still small in 1830, with only 220 members, and although this figure was trebled to around 700 in 1850, the increase was scarcely adequate to cope with the need for railway engineers. The view of the 'Railway Mania' expressed by Samuel Smiles, that; 'no scheme was so mad that it did not find an engineer, so called, ready to endorse it,' undoubtedly represented a widespread opinion at the time, even though Smiles later edited out this and other passages which reflected adversely on the engineering profession.[13] Nevertheless, it was clear that the profession was being encouraged to expand at an unprecedented rate, without any guarantee that the demand for more engineers would be maintained in the future.

In view of this uncertainty about long-term trends, some leading members of the Institution of Civil Engineers were inclined to be cautious.[14] Such Institutional caution was justified in anticipating an end to the railway boom in Britain, but it failed to appreciate the capacity of the profession to expand into a wider world market. A severe set-back in the demand for railway engineers certainly occurred in Britain at the end of the 1840s, with the collapse of the Railway Mania and the revelations of chicanery which had been practised with the shares of railway companies by George Hudson. Thereafter, the need for civil engineers to build British railways fell off sharply, and although there was a continuing growth in demand for mechanical engineers to build locomotives and equipment, which was reflected in the establishment of the Institution of Mechanical Engineers as an independent body in 1847, the profession as a whole was confronted by a surplus of trained talent. Throughout the second half of the nineteenth century, therefore, a steady flow of British engineers was maintained to all parts of the world.

The dominant pattern of this movement had already been established by the response of the railway pioneers to invitations to build railways abroad. Almost from the start of the railway boom, British engineers had been going overseas to supervise railway construction, usually taking teams of British assistant engineers with them as well as British craftsmen and labourers, and generating orders for British iron foundries and machine workshops. The first market for British railway engineering expertise was in Europe. George and Robert Stephenson visited Belgium in 1835 to advise on the first railway in that country, and Robert subsequently built railways in France, Italy, and Norway.[15] Joseph Locke received important commissions for railways in northern France, for which Thomas Brassey acted as his contractor with an 'army' of British navvies.[16] I. K. Brunel was invited to build a railway from Genoa to Turin for the Kingdom of Savoy-Piedmont, and although he eventually pulled out of this project after many frustrations, he and his able Assistant Engineer, Benjamin Herschel Babbage, went on to construct a short stretch of railway north from Florence.[17] William Cubitt was consulted about the Paris-Lyon and Boulogne-Amiens railways: he also did harbour works in Hamburg and provided Berlin with a new water supply.[18] C. B. Vignoles also gave advice on several continental lines, and came to maintain 'a large professional staff' in St. Petersburg.[19]

Where the railway engineers led, others were quick to follow, so that British engineers were ranging across the continent in the mid-nineteenth century. William Handyside was engaged on building an Imperial arsenal in Russia.[20] William Lindley was building railways in northern Germany, and equipping Hamburg and Frankfurt-on-Main with their first modern water works: he also provided a system of sewers for Warsaw.[21] William Husband was erecting the massive steam pumping engines at Haarlemmermeer in the Netherlands for Harveys of Hayle.[22] Aaron Manby was setting up engineering factories in France, to make that country independent of British steam technology.[23] John Haswell was establishing a company in Vienna to build steam locomotives.[24] Philip Taylor had taken his patent for horizontal steam engines to France, where he had founded engineering works at Paris and Marseilles.[25] Alfred S. Jee, a Liverpudlian who had been a pupil of Locke, was killed in Spain in 1858 when an embankment collapsed on the Santander Railway which he was building.[26] And C.E.L. Brown was persuaded in 1851 to join the Sulzer works at Winterthur, where he became 'the Father of Swiss engineering', manufacturing steam engines and electrical equipment.[27] The cumulative impression of all these and similar projects was one of a very significant outward movement from Britain both of capital and of engineering expertise. Not all this expertise was of professional calibre, but the fact that many engineering craftsmen joined in the migration of British talent makes the problem of quantifying the process even more difficult.[28]

After Europe, the next objective of the engineering entrepreneurs who were prepared to go abroad in search of professional commissions was America. The movement of expertise across the Atlantic had already begun before the stimulus of the railways: we have already noted the transfer of steam and textile technologies in the eighteenth century, and the abortive attempts to re-open metal mines in South and Central America in the 1820s. The United States of America was quick to recognise the value of the railway and the steam locomotive, providing one of the first overseas markets for the engines which were turned out in large numbers from Robert Stephenson's engineering company in

Newcastle-on-Tyne from the 1820s onwards.[29] However, in this post-colonial period of American history there was little inclination to rely for long on British expertise and engineers, so that within a very short period American engineers were building their own railways and equipping them with their own locomotives, specially adapted to the conditions of the North American continent.

Canada, on the other hand, remained more receptive to British engineering penetration. The Royal Engineers had been active in the improvement of waterways and other transport systems from the beginning of the nineteenth century, and they continued to play an important part in the development of railways. The Grand Trunk Railway was established under the terms of the Guarantee Act of 1849, whereby the Canadian government undertook to pay interest on the securities of railways over seventy-five miles long - a device also adopted in India and elsewhere to encourage investment in railways. Brassey was appointed principal contractor, and brought over 10,000 navvies for the work. Robert Stephenson designed a large tubular bridge, modelled on his Menai Britannia Bridge, across the St. Lawrence River at Montreal. He visited the site in the spring of 1852, and the 'Victoria Bridge' was built between 1854 and 1859, all the wrought-iron work having been manufactured and partially pre-fabricated in Britain. It was opened by the Prince of Wales in a ceremony which included the installation of an elaborate plaque recording the names of all the bridge engineers.[30] Brassey built the Montreal to Toronto stretch of the railway, providing the first link of a transcontinental route. The Scottish engineer Sandford Fleming, who had been born in Kirkaldy in 1827 and emigrated to Canada in 1845, completed the eastwards link to Halifax in Nova Scotia as the Interprovincial Railway, and went on to survey the route for the Canadian Pacific Railway, built to link British Columbia with the eastern provinces, which was completed in 1885. Fleming received a knighthood and lived on to 1914 as a respected 'statesman-engineer' of his adopted country.[31]

In other parts of the American continents, British engineers made a profound contribution to railway, urban, and water-supply engineering in the second half of the nineteenth century. In 1874, the President of the Institution of Civil Engineers drew attention to three distinguished British engineers operating in South America: J. F. la Trobe Bateman, who undertook water-works in Buenos Aires; Edward Woods, who built railways in Peru, Chile and Argentina; and J. Brunlees, who built railways in Brazil and Uruguay.[32] Others were also active in South America: Robert Francis Fairlie went out to build a railway in Venezuela in 1873, and had previously built narrow gauge lines in Mexico and Brazil.[33] Without any doubt, South America owed much of its rapid economic development in this period to the application of British capital and engineering talent.

A similar range of engineering enterprises dispersed British expertise throughout the Middle East, Africa, and the Far East. In these parts of the world, trade tended to follow the flag, and trade generated the need for engineering works. Engineering thus became an integral part of late nineteenth century British imperialism. In Egypt, for instance, British engineers found both inspiration and important commissions. I. K. Brunel adopted an Egyptian style for his first major work, the Clifton Suspension Bridge in Bristol, and he visited Egypt to recuperate when his health was failing in 1858-9.[34] Robert Stephenson built two tubular bridges for the Alexandria and Cairo Railway, one spanning the Damietta branch of the Nile at Benha and the other the Karrineen Canal at Bir-ket-el-Saba, both being opened to traffic in 1855. Stephenson also examined the Isthmus of Suez for a canal scheme in 1847, but decided that it was not feasible and spoke strong-ly against de Lesseps' French-backed scheme when this was raised in the House of Com-mons in 1857-8 (he was Conservative MP for Whitby at the time), and it seems likely that the condemnation of the leading British civil engineer of the day must have deterred many from investing in the project, which went ahead successfully as a joint French-Egyp-tian enterprise.[35] Nevertheless. British engineers later asserted their influence in Egypt. John Fowler achieved such a thorough knowledge of Egypt from his railway surveys in the 1870s, being consulted by the Khedive, Ismail Pasha, that he was able to advise the British government on the conduct of its Egyptian affairs, for which services he was knighted in 1885.[36] Fowler's business partner, Benjamin Baker, was responsible for a notable piece of colonial annexation, as he was the engineer who shipped Cleopatra's Needle to London. He was subsequently commissioned to design the Aswan Dam, com-pleted in 1902.[37]

Further south, in central and southern Africa, British railway and mining engineers found plenty of scope for their talents in the second half of the nineteenth century, espe-cially with the diamond and gold mining operations and the imperial aggrandisement as-sociated with Cecil Rhodes. The railway engineer Sir Charles Metcalfe (unusually for engineers, he was a baronet by inheritance) had been at Oxford with Rhodes, and went out to southern Africa to undertake large railway building projects including a section of the visionary Cape-to-Cairo line and the great bridge over the Victoria Falls.[38]

India offered even greater career opportunities in British colonial engineering. There were several attempts to launch Indian railway ventures in the 1840s and 1850s, includ-ing the Great Indian Peninsula Railway for which J. J. Berkeley was appointed chief en-gineer in 1849 with the backing of recommendations from Robert Stephenson, Brunel, Cubitt, and other leaders of the profession.[39] This system, eventually containing 1,237 miles of track, was opened in 1853. But it was only after the Mutiny in 1857 and the as-sertion of direct rule by the British Raj that a thorough-going scheme to cover the sub-continent with a network of railways was undertaken. By the end of 1859, eight companies had been formed for the construction of 5,000 miles of track with a capital of £52,500,000. The finance was raised by a government guarantee of 5% on any investment, and by a free gift of land, but in return the government exercised close supervision and slowed down construction by insisting on double track and other conveniences. The guaranteed railways were built on a 5 ft. 6 ins. gauge, but state railways were usually built on a metre gauge. 'The lines themselves were constructed after the best methods then known in Eng-land,' stated an official report in 1908, implying that they were more expensive and bet-ter appointed than was justified by their circumstances.[40]

Many British engineers had a share in this process of railway construction. I. K. Bru-nel was consulted about the East Bengal Railway, but his premature death in 1859 prevented any active involvement, although his nominee W. A. Purdon became Chief Engineer, and his assistant Bradford Leslie went out to India to work on bridges for the project.[41] A. M. Rendel, son of the distinguished harbour engineer J.M. Rendel, advised on the East India Railway and other lines, including the Lansdowne Bridge over the Indus and the Hardinge Bridge over the Ganges.[42] John Brunton, one of another well-known dynasty of British engineers, who is somewhat unusual because he wrote down his engineering recollections for posterity, provides a useful insight into the attitudes and achievements of the colonial engineers of this period. He went to India after serving as Brunel's assistant in the construction of the prefabricated hospital at Renkioi in Turkey for victims of the Crimean war, and on his arrival in India he narrowly missed being caught up in the violence of the Mutiny. He went on to build the Scind Railway and other lines in northern India, recounting many anecdotes of adventures with the natives, whom he regarded with a typical imperial mixture of condescension and impatience.[43]

It was not only railway engineers who found employment in India in this period of Imperial expansion: engineers also came to the country to provide irrigation schemes, water works, and telegraphic services. Systematic construction of irrigation canals, known as 'perennial' canals, to compensate for unreliable rainfall in northern India by channelling water many hundreds of miles from the upper reaches of the Ganges and other rivers, had begun in the 1820s. It was mainly under the direction of military engineers, although not invariably members of the Corps of Royal Engineers. Sir Proby Cautley, for instance, recently described by Joyce Brown as 'the father of the perennial canal system in northern India',[44] had gone out to the sub-continent in 1819 as a young artillery officer, and was directed to canal engineering by the circumstances of the time. Thus, as Brown has observed: 'much of the engineering depended on knowledge gained empirically.' In this respect Cautley was fairly typical of British engineering in general, and to some extent his experience re-traversed the ground already covered by the canal builders of the previous generation in Britain. He remained in India until 1854, except for one furlough of three years in the 1840s, much of which he spent on examining canal works in Britain and on the continent. His greatest work in India was the Ganges Canal, the main trunk of which was almost 900 miles long. He developed a technique of concrete construction in wells sunk to provide foundations, and he was responsible for a large aqueduct at Solani.

Cautley had many assistants and collaborators in this work, including Robert Smith, the Army officer in charge of the Eastern Jumna Canal when Cautley arrived in India to work on the project, Robert Napier and Richard Strachey,[45] although much of the work was supervised by Engineer Sergeants who frequently worked with great dedication in difficult conditions. The irrigation canals were placed under a new Public Works Department in 1854, and became sound financial successes, producing a steady revenue for the government. After retiring to Sydenham in Kent, Cautley continued to give advice on Indian engineering, and became involved in a protracted controversy with Sir John Cotton, a man whose career had corresponded closely to his own as Cotton had performed excellent work as engineer to the great deltaic canals of south east India in Madras.[46] The dispute called in question the efficiency of Cautley's design for the Ganges Canal, and did not enhance the reputations of either of the two men, who had both been knighted for their engineering services to India.[47]

Cautley's own training as an artillery officer had been limited to a few years at Addiscombe Military Seminary, a military school established near Croydon by the East India Company in 1809 in order to train boys for the Indian service, but his anxiety to ensure the supply of personnel trained in civil engineering led him to support the establishment of the Roorkee College of Engineering, opened in 1848.[48] This was the first of four engineering colleges founded in India in the nineteenth century, the others being at Sibpur (Calcutta), Madras, and Poona.[49] These appear to have been designed to attract young men from Britain, but came to provide training for local Indian talent.

The Royal Engineers established a special training college for recruits intending to go out to India at Cooper's Hill, near Windsor, in 1870. It was a response to the post-Mutiny spate of engineering works in India, and provided a three-year course to equip young men for engineering in the sub-continent. Its first President was Colonel Chesney, RE, but when he retired in 1880 the intake was already beginning to falter and students were going to places other than India, so that it ran down until it was closed in 1906.[50] But for thirty years it served the engineering needs of the British Raj in India well, providing in the words of a military officer of the time, 'a better, abler and more gentlemanly set of men than any others available.'[51]

In the field of telegraphic engineering, W. E. Ayrton installed an extensive system in India and developed a successful method for detecting faults in overland wires,[52] while J. U. Bateman-Champain worked on the overland telegraph from Persia to India.[53] In general, the engineers brought a western industrialising influence to the sub-continent which was rapidly assimilated and adapted by the Indian population. However, neither the military nor the civilian engineers made permanent homes in India, although many, like Cautley, did long spells of duty. White settlement was minimal throughout the period of British rule, which made the eventual withdrawal of the British Raj comparatively painless. India was thus left to enjoy the fruits of British imperial engineering without having to accommodate the bitter aftermath of colonial settlement.

Elsewhere in the Far East, the social effects of British engineering were even more attenuated by the presence of a large native population without any permanent European settlement. Large strategic installations at Singapore and Hong Kong encouraged trade and generated vigorous urbanisation, which in turn led to a need for civic buildings and public works such as water supply and waste removal. British engineers also figured in railway building enterprises throughout the Far East. As early as 1876, a British contractor, John Dixon, working for Ransome and Rapier of Ipswich, built the first railway in China, a ten-mile line of 2 ft. 6 ins. gauge from Shanghai to Woosung, equipping it with two locomotives made by Ransomes.[54] However, the aloofness of Chinese imperial bureaucracy to western technology, together with the legacy of the Opium Wars and the Taiping Rebellion, made China less attractive to British engineers than India or even Japan, and the main work of railway building in China came after the Revolution of 1911 led by Sun Yat Sen.[55]

Even in Japan, after this country had been re-opened to western influences, British engineering expertise was readily received, although it is indicative of the later development of Japan that the greatest priority was given to securing first-class university instruction in engineering. To this end, Japanese agents visited Britain to examine its training procedures for engineers, and then set up a college on the Western model, recruiting a team of British engineers and scientists to staff it.[56] W. E. Ayrton was appointed Professor of physics and telegraphy in Tokyo in 1873, and amongst other British engineers who later rose to eminence in the profession was J. A. Ewing, who was Professor of mechanical engineering in Tokyo from 1878 to 1883, before returning to a distinguished university career in Dundee, Cambridge, and Edinburgh, for which he was eventually knighted as Sir Alfred Ewing.[57]

Australia probably provides the neatest case-study of British colonial engineering. It is a study of the transfer and ready adoption of British techniques, so that they swiftly took root and began to produce an indigenous engineering tradition.[58] Right from the beginning of European settlement, in 1788, there had been a need for engineering works, and these had been provided as efficiently as possible by military officers such as Alt, the first Surveyor-General who was responsible for building the original roads, small mills, and reservoirs; by men with engineering knowledge amongst the convicts, like F. H. Greenway, the architect of some of Sydney's finest surviving public works; and by engineers who could be persuaded to come out to the colony, like John Busby, who claimed to have served under John Rennie and who arrived in Sydney in 1824 in order to improve the water supply by the construction of the conduit known as 'Busby's Bore'.[59] Mechanical skills were at a premium amongst immigrants, both voluntary and involuntary, so that men like John Dickson, who arrived in 1813 with the parts of the first Australian steam engine in his luggage, and David Lennox, who arrived in 1832 after working under Telford and who constructed many fine bridges in the colony, were sure of a warm welcome. Most of these men were mechanics rather than professional engineers, but the new settlements in Australia offered them opportunities similar to those which had converted mechanics into professional engineers in Britain in the eighteenth century, and in the same way professional interests began to grow in Australia.

The problems facing this incipient engineering profession in Australia at the beginning of the nineteenth century were those which faced all white settlement on the continent - the sheer size of the place, its natural obstacles, its lack of capital resources, and its distance from Europe.[60] These problems began to be overcome in the second half of the century, under the stimulus of the succession of gold rushes which greatly increased the rate of immigration and which brought in their wake a flow of mechanical expertise associated with Cornish mining techniques. Cornish miners moved all over the world in pursuit of metallic ores, taking with them distinctive methods of working lodes of metals and of processing ores and, perhaps most pervasive and important, knowledge of the versatile Cornish engine. This British prime mover was universally adopted in the second half of the nineteenth century where a powerful and reliable pump was required to clear deep mines of water together with a winding engine to raise the ore and the miners. Their derelict engine houses can now be found all over the world in deserted mining areas, and Australia has an ample share. Nor were the Cornish miners alone in spreading British metal-working expertise. In South Australia and elsewhere, Welsh communities of copper smelters settled and applied their skills to the benefit of themselves and to the prosperity of the Australian colonies.[61]

The boom in mining which followed the first Gold Rush coincided with the dispersal of British railway expertise in the 1850s and thereafter. Thomas Barker had arrived in Australia as an apprentice to John Dickson as early as 1813, but he survived until 1875 and lived to become an important promoter of New South Wales railways. Other states depended more directly on British engineers. In 1851, for instance, South Australia commissioned B. H. Babbage, who had been Brunel's assistant on his Italian railway ventures, to build its first railway. Babbage settled in the colony and went on to do important exploration and surveying work in the hinterland.[62] Peter Russell, whose family had arrived in Hobart from Scotland in 1832 and then moved to Sydney, made a fortune from his iron foundry in the manufacture of railway lines and other railway equipment. He also moved into shipbuilding, when the need for coastal steam packets was increasing rapidly in Australia. Other engineers, like John Ridley who had emigrated to South Australia from Northumberland, recognised the need for farm machines and established successful firms to produce them.[63] Others, again, were called upon to provide urban services, such as James Blackburn, who constructed Melbourne's first water supply system, and A. K. Smith, who provided its gas works. For its sewage disposal problems, Melbourne consulted the British engineer James Mansergh in 1889. In harbour works also, British engineers were in considerable demand: John Fowler advised on the Cockatoo Island docks in Sydney, as well as on New South Wales railway schemes; Sir John Coode, the outstanding harbour expert of the late nineteenth century, advised on various Australian port installations; and C. Y. O'Connor, an Irishman who came via New Zealand to Western Australia, was appointed Engineer in Chief to that state in 1891.[64]

By the end of the nineteenth century, Australian engineering was well established in all of its main branches, and ready to undertake large-scale projects such as the Coolgardie water supply scheme, engineered by O'Connor, and the major developments in metal working promoted by the Broken Hill Proprietary Company, which led to a significant increase in the recovery of workable metals from their ores. These innovations were significant also because they involved a departure from reliance upon British engineering expertise, the Broken Hill entrepreneurs preferring to turn to German and American metallurgists for assistance in the key processes.[65]

The establishment of schools of engineering at Melbourne and Sydney Universities helped to promote the indigenous Australian profession of engineering, even though they still turned largely towards British experts as teachers. W. H. Warren, for instance, had served his apprenticeship on the London and North Western Railway before migrating in 1881 to a post in Sydney Department of Public Works, from which he moved into the University and became Professor of Engineering and first President of the Institution of Engineers, Australia, when this was formed in 1919 by a federation of local institutions of which the oldest, in New South Wales, dated back to 1870.[66] W. C. Kernot, the other outstanding pioneer of engineering education in Australia, has been taken out to Victoria at the age of six in 1851. He became the first qualified engineer to be produced by the University of Melbourne in 1866, and went on to become the first Professor of Engineering at Melbourne in 1883.[67] Thus, by the First World War, Australia had fully outgrown its colonial dependence on British engineering. Even though experts from overseas were still consulted in major schemes such as the Sydney Harbour Bridge and the Snowy Mountains Project, the consultation was now between equals in a world-wide community of engineers.

It was not only in Australia that the First World War effectively marked the termination of British supremacy in world industrialisation. For all practical purposes, the War represented the end of the unchallenged supremacy of the railways, with the internal combustion engine leading a dramatic revival of road transport; and with the end of the supremacy of the reciprocating steam engine as the main industrial prime-mover, the age of the steam turbine and electric motor had dawned. The War also brought the beginning of the end of the traditional British Empire, although it remained to loom large in red colouring on Mercator's map projection of the world for several decades thereafter. In all these respects, British engineering experienced a diminution in overseas demand. It had served the developing world well in the nineteenth century, but after 1919 it had little to offer in exciting technological innovations and it found itself in competition in many of its traditional colonial markets both with more dynamic competitors from countries with better systems of technical instruction, and from indigenous engineering industries which had frequently been set up by British engineers and which had then developed more expertise than their British mentors. Australian experience illustrates these changes neatly. Steam engines from firms like Tangyes and Marshalls were being replaced by American and Australian models, or superseded altogether by petrol and diesel engines introduced from America and elsewhere, and a similar process of replacement was taking place in agricultural machinery, machine tools, and domestic equipment. British engineers were still consulted on important occasions, especially in civil engineering, but there was little continuing demand for British engineering experience. The time had passed when there was a Cornish 'Cousin Jack' at the bottom of every hole in the world where metal was being extracted, and when somebody answered to 'Scotty' in the engine room of every steam ship. The great days of British engineering superiority were over.

Just as the 'swarming' of British engineers overseas in the second half of the nineteenth century was the result of hundreds of individual decisions rather than of a conscious strategy devised by government or any other central authority, imperialist or otherwise, so the impression made by these engineers on the societies to which they went was very uneven. The initial motivation had been a technological imbalance, first between Britain and the other Western nations, and then between the Western nations and the rest of the world. In the first of these phases, British expertise was rapidly assimilated by the nations of Western Europe and the United States, and with few exceptions there was no long-term settlement of British engineers. In the second phase, however, the transfer of western expertise to the comparatively undeveloped parts of the world involved a more enduring contribution by British engineers, either because emerging 'Western' style nations like Canada and Australia were anxious to take advantage of British skills over a prolonged period, or because imperial possessions such as India were subjected to an intensive programme of development under the direction of military and civil engineers. In the case of Japan, a non-Western nation demonstrated a remarkable aptitude for acquiring Western expertise at its source by sending its own young men to be trained as British engineers, and by persuading British engineers to teach them in Japanese colleges. So the impact of British engineers varied greatly, depending both on the receptivity of each country and territory concerned, and on its ability to assimilate skills in Western technology. European nations and the USA dispensed fairly quickly with British assistance. On the other hand, Canada, Australia and Japan built steadily on British experience and liberated themselves from reliance upon it. At the extreme pole of reliance, India became heavily dependent on British engineers, while receiving little encouragement to develop its own resources of talent until the last days of the Raj.

Although the spread of British engineers abroad was not deliberately conceived as an instrument of Empire, therefore, its deepest and most lasting effects were felt in the new nineteenth century imperialism. It was certainly not 'colonialist' in an exploitative or pejorative sense, but British engineering did make a creative contribution to 'systematic colonisation' in places like Australia where it was put at the service of settlers in developing new territories. Even in India, where engineering developments were closely associated with the need to maintain the political power of the British Raj, the engineering achievements in railway building and irrigation works were quickly assimilated into the way of life of the native population and stimulated a general increase in economic activity. The impact of the engineers was thus greatest in the British Empire, but their influence survived the dismemberment of the Empire and equipped the new nations which emerged from it with a sound infrastructure of transport systems and urban services.

In the last resort, the dispersal abroad of British engineering was limited in time by the unusual circumstances of Britain's virtually unchallenged leadership in world engineering expertise in the second half of the nineteenth century. It derived its initial impetus from steam technology and the boom in railway construction, and was maintained by the gold rushes in America, Africa, and Australia, and the subsequent demand for metal mining expertise and the refinements of steam technology in which Britain led the world. It was also promoted by the great expansion of urban engineering, for water works, sewage and waste disposal, gas works, tramways, harbours, and associated developments. So the diaspora reached its peak between 1850 and 1914, and thereafter subsided, leaving spectacular monuments to the high tide of British engineering achievement in many parts of the world. Its very success in spreading engineering expertise throughout industrialising societies removed the most powerful incentive to engineering mobility. Of course, the demand for outstanding engineers remained great, and many ordinary engineers from all the developed countries found opportunities to travel and to use their talents in foreign countries. But since 1914 no single country has dominated world engineering in a way that could encourage a repetition of the remarkable out-pouring of talent which occurred from Britain in the nineteenth century. And it seems unlikely, moreover, that future circumstances will re-create the technological imbalance between one country and the rest of the world which generated this extraordinary process.

Notes

1. Charles Labeyle was the son of a French Huguenot who had settled in Switzerland: he had come to England as a young man around 1720, and was appointed engineer for the Westminster Bridge in 1738. See R. J. B. Walker, *Old Westminster Bridge - The Bridge of Fools*, David and Charles, Newton Abbot, 1979, 81-3.

2. L. T. C. Rolt, *Thomas Newcomen - The prehistory of the steam engine*, London, 1963, gives a succinct account of the spread of the Newcomen engine overseas in the early decades of the eighteenth century. For the background to the Dannemora engine of 1727, see Svante Lindqvist, 'The work of Martin Triewald in England, in *Transactions of the Newcomen Society*, 50, 1978-79, 165-172, and the same

author's *Technology on Trial - The Introduction of Steam Power Technology into Sweden, 1715-1736*, Uppsala, 1984.

3. A recent study of this technological transmission is David J. Jeremy, *Transatlantic Industrial Revolution: the diffusion of textile technologies between Britain and America, 1790-1830s*, MIT Press, Cambridge, Mass. 1981.

4. John Smeaton's *Diary* of his journey to the Low Countries, published from the original manuscript in the Library of Trinity House, London, with an introduction by Arthur Titley, Newcomen Society, London, 1938.

5. Sir Alexander Gibb, *The Story of Telford - the rise of civil engineering*, London, 1935.

6. See R. F. Leggett: 'The Jones Falls Dam on the Rideau Canal' in *Trans. Newcomen Soc.*, 31, 1957-58, 208-218.

7. Richard Shelton Kirby, 'William Weston and his contribution to early American Engineering' in *Trans. Newcomen Soc.*, 16, 1955-6, 111-128.

8. Carroll W. Pursell, Jr., *Early Stationary Steam Engines in America - a study in the migration of a technology*, Smithsonian Institution Press, Washington DC, 1969. See also William Nelson: 'Josiah Hornblower and the first steam engine in America', in *Proc. New Jersey Hist. Soc.* 2nd ser., 7, 1883, 177-247.

9. Edward C. Carter: *The Virginian Journal of Benjamin Henry Latrobe 1795-1798*, Maryland Historical Society, 1977, esp. viii.

10. Trevithick spent the decade 1816-1827 in Central America. See H. W. Dickinson and Arthur Titley, *Richard Trevithick - the engineer and the man*, Oxford, 1934, esp. Chapter 5 - 'The Great Adventure'.

11. Robert Stephenson was in Venezuela and Colombia from 1824 to 1827: he met Trevithick as a 'tattered castaway' in Cartagena and gave him £50 to help him home. See L. T. C. Rolt, *George and Robert Stephenson - The Railway Revolution*, London, 1960, esp. Chapter. 6, 'The Mines of Santa Ana'.

12. W. O. Henderson, *The Industrialisation of Europe 1780-1914*, London, 1969, 51.

13. S. Smiles, *Life of Stephenson*, 3rd edition, London, 1857.

14. See for instance J. Walker, Presidential Address, *Minutes of Proc. Inst. Civil Engs.*, 1, 1841, 25-6 .

15. See Rolt, *Stephensons*.

16. Locke has not been so well blessed by good biographers as his contemporaries I. K. Brunel and Robert Stephenson. But see N. W. Webster, *Joseph Locke - Railway Revolutionary*, London, 1970.

17. See R. A. Buchanan, 'The Overseas Projects of I. K. Brunel', Presidential Address, *Trans. Newcomen Soc.*, 54, 1982-83, 145-166.

18. Summarised from the *Dictionary of National Biography (DNB)* entry on Cubitt.

19. Summarised from the *DNB* entry on Vignoles. See also K. H. Vignoles, *Charles Blacker Vignoles - Romantic Engineer*, Cambridge, 1982. Vignoles' major commission in Russia was the multiple-suspension bridge over the River Dneiper at Kiev, built between 1846 and 1853.

20. William Handyside (1793-1850): see entry in *DNB*. There had been a trickle of British expertise to Russia ever since the visit of Peter the Great to Western Europe in 1697-8.

21. William Lindley (1808-1900): see obituary notice in *Proc. Inst. Civil Engs*, 142, 1899-1900, 363-370. See also the feature in *The Times*, 23 May 1983: 'How Britain helped the Polish underground.'

22. William Husband (1822-1887): see entry in *DNB*.

23. Aaron Manby (1776-1850): see entry in *DNB*.

24. John Haswell (1872-1897): see F. J. G. Haut, 'The Centenary of the Semmering Railway and its Locomotives, *Trans. Newcomen Soc.*, 27, 1949-51, 19-29, esp. 27.

25. Philip Taylor (1786-1870), see entry in *DNB*, also H. W. Dickinson and A. A. Gomme, 'Some British contributions to continental technology (1600-1850), in *Acts du VI Congres International d'Histoire des Sciences*, Amsterdam, 1950, I, 307-323.

26. Alfred Stanistree Jee (1816-1858), see obituary notice in *Proc. Inst. Civil Engs.*, 18, 1858, 193-6.

27. S. B. Saul, 'The Nature and Diffusion of Technology', in A. J. Youngson (ed.), *Economic Development in the Long Run*, London, 1972, 36-61, esp. 51.

28. For the general course of British initiative and continental emulation, see David S. Landes, *The Unbound Prometheus*, Cambridge, 1970. Landes estimates that 'At mid-century ... continental Europe was still about a generation behind Britain in industrial development' (187).

29. Rolt, *Stephensons*, (Pelican edition, 235) lists six USA companies provided with two locomotives: the Mohawk and Hudson, the Newcastle and Frenchtown, the Baltimore and Susquehanna, the Saratoga and Schenectady, the Charlestown and Columbia, and the Camden and Amboy. The first locomotive supplied to the Camden and Amboy in 1831, the *John Bull*, survives in the National Museum of American History: see John H. White, Jr., *The John Bull - 150 Years a Locomotive*, Washington DC, 1981.

30. Bruce Sinclair, 'Canadian Technology: British Traditions and American Influences', in *Technology and Culture*, 1979, 108-123

31. John W. Abrams, 'Sandford Fleming in Canada', *Trans. Newcomen Soc.*, 49, 1977-78, 133-137. This paper by the late Professor Abrams was a contribution to the joint Newcomen Society - ICOHTEC Symposium on 'Scottish Engineers and Engineering' organised at the Summer Meeting of the Newcomen Society in Stirling in August 1977.

32. Thomas E. Harrison in his Presidential Address: *Proc. Inst. Civil Engs.*, 37, 1874, 225-6.

33. Robert Francis Fairlie (1831-1885), see entry in *DNB*; also P. C. Dewhurst 'The Fairlie Locomotive, Pt. I', in *Trans. Newcomen Soc.*, 34, 1961-2, 105-132; and Pt. II, 1966-7, 1-34.

34. Discussed in L. T. C. Rolt, *Isambard Kingdom Brunel*, London, 1957.

35. Rolt, *Stephensons*, (Pelican edition), 326-7.

36. Thomas Mackay, *The Life of Sir John Fowler*, London, 1900, esp. Chapter X - 'Egypt'. In the same year that Fowler received his knighthood, 1885, General Charles George Gordon, a prominent member of the Corps of Royal Engineers, became a hero of British imperialism with his death at Khartoum.

37. Benjamin Baker (1840-1907), see entry in *DNB*.

38. Sir Charles Metcalfe (1853-1928), see entry in *DNB*.

39. J. N. Westwood, *Railways of India*, Newton Abbot, 1974; P.S.A. Berridge, *Coupling to the Khyber*, New York, 1969; J. Hurd, 'Railways' in *The Cambridge Economic History of India*, 2, c.1757-c.1970, Cambridge, 1980. For J. J. Berkeley (1819-1862), see entry in *DNB*.

40. *The Imperial Gazetteer of India: the Indian Empire III Economic*, Oxford, 1908, Chapter VII 'Railways and Roads, 383.

41. See R. A. Buchanan, 'The overseas projects of I. K. Brunel' (note 24 above). Bradford Leslie undertook major bridge works in India: see his paper 'Account of the Bridge over the Gorai River, in *Proc. Inst. Civil Engs.*, 34, 1872, 1-42.

42. Alexander Meadows Rendel (1829-1918), see entry in *DNB* (1912-1921).

43. *John Brunton's Book*, introduced by J. H. Clapham. Cambridge, 1939.

44. Joyce Brown, 'Sir Proby Cautley (1802-1871), a Pioneer of Indian Irrigation', in A. Rupert Hall and Norman Smith (eds), *History of Technology*, 3, 1978, 35-89.

45. Joyce Brown, (no. 44. above), 43, 53, and 59.

46. Arthur Thomas Cotton (1803-1899), see entry in *DNB* (Su).

47. Joyce Brown, (no. 44 above) gives a careful account of the controversy, 77-83.

48. Joyce Brown, (no. 44 above), 84.

49. *Imperial Gazetteer*, (no. 40 above), IV, (1909), 321

50. See Whitworth Porter, *History of the Corps of Royal Engineers*, London, 1889, 2, 172-7.

51. J. G. P. Cameron, *A Short History of the Royal Indian Engineering College Coopers Hill*, issued by the Coopers Hill Society for private circulation, London, 1960, 11: Cameron is here quoting an article in 'an Indian journal' in 1878 by a Military Officer in charge of the construction of the North Bengal State Railway.

52. W. E. Ayrton (1847-1908), see entry in *DNB* (1901-1950).

53. John Underwood Bateman-Champain (1835-1887), see entry in *DNB* (Supp.).

54. See *Engineering* 14 July 1876, 29: I am grateful to my colleague Dr. D. Brooke for this reference.

55. See Kenneth Cantlie, *The Railways of China*, London, 1981, also the same author's paper 'The Chinese 4-8-4 Locomotive', *Trans. Newcomen Soc.*, 54, 1982-83, 127-148. For general background, see Ssu-yn Teng and J. K. Fairbank, *China's Response to the West*, Harvard, 1954.

56. See W. H. Brock, 'The Japanese Connexion: engineering in Tokyo'. London and Glasgow at the end of the nineteenth century' in *British Jnl. Hist. Science*, 14, Pt.3, No. 48, November 1981, 227-243.

57. Sir Alfred Ewing, *An Engineer's Outlook*, London, 1933; A. W. Ewing, *The Man of Room 40 - The Life of Sir Alfred Ewing*, London, 1939. For another example of British expertise in Japan, see Richard F. Trevithick, 'Locomotive Building in Japan, in *Proc. Inst. Mech. Engs.*, 1895, 298-307.

58. See R. A. Buchanan, 'The British contribution to Australian engineering', in *Historical Studies*, University of Melbourne, April 1983, 401-419.

59. See A. H. Corbett, *The Institution of Engineers, Australia*, Sydney, 1973, which has a useful first chapter on engineering developments up to 1919, when the Institution was founded. For biographical information on Australian engineers, the *Australian Dictionary of Biography* (*ADB*) is invaluable.

60. See G. Blainey, *The Tyranny of Distance*, Melbourne, 1966.

61. D. A. Cumming and G. Moxham, *They built South Australia - Engineers, Technicians, Manufacturers, Contractors and their Work*, published by the authors, Adelaide, 1986.

62. See the entry for B. H. Babbage in *ADB*. Also R. A. Buchanan, 'The overseas projects of I. K. Brunel' (no. 17 above), for Babbage's involvement in the Italian railway schemes.

63. See L. J. Jones 'The early history of mechanical harvesting', in A. Rupert Hall and Norman Smith (eds), *History of Technology*, 4, 1979, 101-148. Also L. J. Jones, 'Wind, water and muscle-powered flourmills in early South Australia', *Trans. Newcomen Soc.*, 53 1981-82, 97-118,

64. For biographical references, see entries in *ADB*. Charles Yelverton O'Connor (1843-1902) does not appear in the nineteenth century volumes of the *ADB*, but see, 'The origins of the eastern Goldfields water scheme' by F. Alexander *et al.*, (1954), National Library of Australia, N994.1041.ALE.

65. See G. Blainey, *The Rush that never ended - A history of Australian mining*, Melbourne University Press, 1963.

66. For W. H. Warren, see entry in *ADB*, and also A. H. Corbett, no. 59 above.

67. For W. C. Kernot, see entry in *ADB*.

Chapter Nine

Engineering Education and Training

When William Charles Kernot received the new Certificate of Civil Engineering at the University of Melbourne in 1866, it could be claimed that he was the first academically qualified engineer to be produced by Australia. He went on to become the first Professor of Engineering at Melbourne, and one of the pioneers of academic training for engineers in the English-speaking world. As such he represented a decisive shift within the engineering profession towards the recognition of the value of theoretical expertise obtained through a university discipline. But the fact that this shift was not universally welcome amongst senior engineers is demonstrated by the way in which Kernot was received in his first post. After qualifying at the university, he was employed for a time on the Victorian State Railways under Thomas Higinbotham, an Irish-born railway engineer of the old school who had been employed on railway works in Britain before emigrating to Australia in 1857, to become Engineer-in-Chief of Victorian Railways. While initiating him into railway work, Kernot recalled that Higinbotham 'never missed an opportunity of impressing upon me the uselessness and undesirability of university training for engineers'.[1] Thus the traditional attitudes towards engineering training, which had served the profession well for a couple of generations, came into incomprehending conjunction with the innovations which were transforming engineering attitudes on professional training in the second half of the nineteenth century. It was a conjunction which was repeated on many occasions and at many levels as the profession expanded to meet the new demands for engineering expertise in this period.

There have been several studies of the development of scientific and technological education in nineteenth century Britain, so that it is not necessary to rehearse this theme here.[2] No previous attempt has been made, however, to see these developments from the point of view of the engineering community. As a result, an important dimension has been omitted from the historical interpretation of the social impact of science in this period. In particular, inadequate attention has been given to the significant 'supply push' which contributed to the transformation of the engineering profession in addition to the 'demand pull' provided by increasing national provision for technological education. We will be concerned here with the former rather than the latter. The emphasis will be on the changing attitudes of the engineers, rather than on the content of scientific theory. Three stages will be distinguished in the process by which engineering undertook its own transformation: first, the preliminary perception of a need for technical training and the staking out of the ground for possible development; second, the establishment of firm foundations for theoretical instruction with the recognition by some British universities and other educational bodies of an agreed corpus of such knowledge; and third, the wide-spread approval of a comprehensive programme of higher education in engineering incorporating extensive laboratory work and research. It will be necessary to concentrate attention on the adoption of engineering by university-type institutions, although occasional reference will be made to other developments. For all practical purposes, the transformation was complete in Britain by 1914, so that the emergence of a systematic, theoretically-based pattern of engineering education can be regarded as having been achieved by that date.

The engineering community was still quite small in the middle of the nineteenth century, although it had grown rapidly since the establishment of the Institution of Civil Engineers in 1818 and was to continue to grow for the following century. The combined membership of the Civils and the Institution of Mechanical Engineers was just under one thousand in 1850,[3] and virtually all these 'professional' engineers had acquired their skills by a process of pupilage in the office of an existing engineer, in a manner derived directly from time-hallowed experience of the practical training of apprenticeship. The only significant exceptions were the few who had received military instruction as Royal Engineers at the Royal Military Academy, Woolwich, or in the engineering establishment set up at Chatham in the Napoleonic Wars, and an even smaller number who had been trained in the French *écoles* or elsewhere overseas.[4] The general pattern of British engineering education at this time was thus overwhelmingly practical, with experience in the field, at the drawing board, and at the work-bench counting as much more important than any specific form of theoretical instruction and examination.

Nevertheless, British engineers were not entirely lacking in theoretical knowledge before 1850. Close relationships had existed between many leading engineers and the natural philosophers of the eighteenth century. John Smeaton had presented an outstanding paper on wind and water power to the Royal Society of which he was a member in 1759, and although it could be argued that he provided an initiative in the skilful use of models for testing performances rather than in the elaboration of any striking new theory, his analysis was of great practical value in helping millwrights to improve the characteristics of their designs.[5] James Watt, also a member of the Royal Society, had benefited from Joseph Black's pioneering studies of heat at Glasgow before inventing the separate condenser for the steam engine in 1769, and had become a member of the distinguished group of provincial men of science which gathered in Birmingham as the Lunar Society.[6] The works of French and German authors on hydrostatics, hydrodynamics, and metallurgy, were widely circulated, and there were lively exchanges with continental chemists like Berthollet and Lavoisier.[7] The strength of materials was in its infancy as a theoretical study in the eighteenth century, and most engineers had only a rudimentary knowledge of the principles of tension, compression, friction and fatigue. Not until the experimental work on cast- and wrought-iron railway bridges in the middle of the nineteenth century did it begin to be recognised as a subject for serious scientific investigation. The related study of soil science was even slower in emerging, although here again engineers acquired considerable practical experience, often related to the advances in geology, which numbered canal and mining engineers like William Smith amongst its British pioneers.[8] A modern study has calculated that Telford's design for a single cast-iron arch to replace the old London Bridge in 1801 was not seriously at fault in its assessment of rib stress and the strength of the abutments.[9]

This relationship between engineers and men of science developed further in the nineteenth century. Engineering experience stimulated the emergence of thermodynamic theory.[10] The Brunels, both father and son, were friends of Michael Faraday, and I. K. Brunel was amongst the first people to recognise the value of the electric telegraph as a means of railway signalling.[11] Several engineers took an active part in the early days of the British Association, which provided a forum for the famous altercation between I. K. Brunel and Dionysius Lardner about the feasibility of trans-oceanic steam navigation.[12] Brunel also entered into controversy with Lardner and William Buckland about the geological soundness of the Box Tunnel; and when, like other engineers, he became keenly interested in the reasons for the failure of iron railway bridges in the 1840s, he undertook lengthy calculations and tests to solve the problem.[13]

Of all the British engineers of the mid-nineteenth century, it might have been expected that I. K. Brunel would have shown a greater interest in science and theoretical knowledge than his contemporaries. He was, after all, the son of a distinguished French engineer who had benefited from a theoretical training, and he had himself enjoyed the opportunity of some years of formal tuition in mechanical subjects in France. His Drawing Books and Calculation Books, moreover, demonstrate an excellent grasp of mathematical principles. But as far as theoretical training was concerned - as distinct from the theoretical knowledge gained from experience - Brunel possessed the caution amounting to suspicion which was typical of mid-century British engineering. He expressed himself explicitly on the subject in a letter to a young man hoping to become an engineer:

'I must caution you strongly against studying *practical* mechanics among French authors - take them for abstract science and study their statics dynamics geometry etc. etc. to your heart's content but never even read any of their works on mechanics any more than you would search their modern authors for religious principles. A few hours spent in a blacksmiths and wheelwrights' shop will teach you more practical mechanics - read *English* books for practice - There is little enough to be learnt in them but you will not have to unlearn that little.'[14]

Brunel's views of the nature of engineering thus differed less than might have been expected from those of his colleagues in the profession. His conventional, free-trade, *laissez-faire* views were offended in 1848 when the government appointed a Royal Commission to investigate the construction of railway bridges following the disastrous collapse of Robert Stephenson's Dee Bridge at Chester, a composite structure of cast-iron beams and wrought-iron supporting members derived from what were admittedly very hazy notions of the behaviour of these metals under heavy loads. Brunel described the inquiry as 'the commission for stopping further improvements in bridge building'[15] and applied himself with characteristic vigour to denying the possibility that the commission could do anything by regulation to improve a matter which was essentially a problem for engineers to solve themselves. However old-fashioned this view might appear to a generation accustomed to a high degree of governmental regulation of engineering standards, the most important consequence of the Chester tragedy was, as Brunel anticipated, that the engineers learnt valuable lessons from it. Stephenson and Brunel both abandoned cast-iron as a bridge-building material for railway works, and embarked on elaborate test procedures for wrought-iron structures. The most famous of these were the model tests carried out for Stephenson by William Fairbairn and the mathematician Eaton Hodgkinson between 1845 and 1848,[16] which resulted in the novel tubular girder design used to such brilliant effect in the Britannia Bridge over the Menai Straits. This has rightly been recognised as a highly significant break-through in scientific engineering, involving the application of metallurgical and strength of materials theory as well as pioneering the use of large-scale models to simulate performance under load.

While these practical foundations of scientific engineering were being laid in Britain, another area of engineering competence was responding positively to the need for a systematic and theoretical treatment, and Brunel was involved in this at the same time as he was improving techniques of bridge construction. This related field was ship building and, in particular, the design of ship hulls to achieve maximum efficiency in movement through water. Brunel's involvement in this subject arose from his practical need to design the best possible hull shapes for his first two steam ships, the *Great Western*, launched in 1837, and the *Great Britain*, launched in 1843. In the course of this work he met John Scott Russell and became familiar with the 'wave line' theory, which Russell had tested with models in tanks and on canals.[17] As a result of this meeting of minds in a common pursuit of sound theoretical knowledge, Russell and Brunel collaborated on the design of Brunel's third and final ship, the giant *Great Eastern*. The partnership turned out to be tortured and disastrous in personal terms, but technically it was a remarkable success, the vessel - the largest built until the twentieth century - fulfilling all the expectations of her designers. Russell went on to found the Institution of Naval Architects and to write the monumental and definitive work combining practical and theoretical knowledge of ship design, *The Modern System of Naval Architecture for Commerce and War*, published in three large volumes in 1865.[18] It was another of Brunel's Assistant Engineers, however, who really converted ship model-testing into a systematic scientific procedure. This was William Froude, who had joined Brunel to work on the Bristol and Exeter Railway in the 1840s, having graduated from Oxford with a first class degree in mathematics in 1832 and having served with H. R. Palmer on the South Eastern Railway. Between 1844 and 1859 he withdrew from the profession for family reasons, although he did do some calculations at Brunel's request on the shape of the *Great Eastern* in 1856. From 1859 until his death twenty years later, Froude became involved in devising procedures for accurately testing the performance of ship hulls by using scale models in large tanks, for which work he received several honours including the Royal Medal of the Royal Society in 1876.[19]

There is thus substantial evidence of an increasingly prominent scientific dimension to British engineering practice in the middle decades of the nineteenth century, in the sense that it was acquiring a more systematic organisation and a distinctive theoretical basis. But there was little corresponding recognition of any need to modify the training of engineers in order to incorporate more theoretical knowledge, either in the profession or the universities. The views of members of the Institution of Civil Engineers were expressed by their President, James Walker, in his Presidential address of 1841. There was an uneasy sense that the development of academic engineering posed a threat to traditional methods of instruction, and threatened to swamp the market with a flood of theoretically qualified engineers for whom there would not be sufficient jobs. Walker congratulated the Institution on its growth, but went on to express anxiety about the likely over-supply of engineers:

'Is then the demand for professional gentlemen likely to *increase*? Is it not likely rather to *decrease*? Now certainly the number of Engineers or Students for Engineering is increasing. If we look at the number of students in the classes for Civil Engineering at the different Universities and Academies, the Universities of Edinburgh and Durham; King's College, University College, and the College for Civil Engineers in London; we are led to ask, will this country find employment for all these? I freely confess that I doubt it.'[20]

Of the courses mentioned by Walker, only those at University College and King's College, London, had any real permanence. The founders of University College had intended to set up a department of engineering at the outset in 1828, and actually offered the first chair of 'Engineering and the Application of Mechanical Philosophy to the Arts' to J. Millington, who was already a professor at the Royal Institution and had been associated with Birkbeck in setting up Mechanics' Institutes. But he resigned soon after the opening of the College on account of disagreements about his salary. Thereafter, for several years, only occasional lectures on engineering were offered, despite intermittent negotiations with the Institution of Civil Engineers regarding a new appointment, until the chair of civil engineering was founded in 1841. The railway engineer Charles Vignoles became the first holder of the chair, from 1841 to 1845, when he resigned owing to increased pressure of professional business.[21] Subsequently, Eaton Hodgkinson became Professor of mechanical principles of engineering in 1846, but the chair lapsed with his death in 1861. There was a revival in 1867, when Fleeming Jenkin briefly held a new chair in civil engineering, and then in 1874, when A. B. W. Kennedy was appointed to the chair of applied mathematics and mechanics. As we will observe shortly, it was Kennedy who put UCL in the forefront of British engineering education.[22]

Meanwhile, King's College London had appointed Charles Wheatstone as Professor of experimental philosophy in 1834, and in 1838 a department of civil engineering was opened offering courses in mathematics, mechanics, chemistry, geology, electricity, machine-drawing and surveying. There were thirty-one students in the first year, fifty in 1839, and fifty-eight in 1840. In the latter year, a workshop was opened and architecture was added to the syllabus, and William Hosking, described by the historian of the College as 'one of the leading hydraulic architects and railway engineers of the day'[23] accepted the chair of 'the arts and construction'. There was a fall in student numbers to thirty-three in 1844, but they rose again with the boom in railway construction and remained substantial thereafter.

Walker had spoken of classes for civil engineering also at the London College of Civil Engineers, and at the universities of Edinburgh and Durham. The London 'college' was opened in Putney in 1834, its staff including the gas engineer Samuel Clegg. Its aims were professional rather than simply educational, and in 1857 it became the Society of Engineers, which still survives as a professional and social club.[24] Of the two provincial universities mentioned, the courses at Edinburgh seem to have been experimental and short-lived, as serious engineering education did not begin there until the 1850s.[25] The Durham development was more ambitious, although it also proved to be short-lived, demonstrating in microcosm the problems of engineering education in the first half of the nineteenth century. The 'Northern University' had been established at Durham in 1832, teaching predominantly arts students but with a strong infusion of mathematics and chemistry. The Professor of mathematics, Cambridge mathematician the Rev. Temple Chevallier, and the lecturer in chemistry, James F. W. Johnston, took the initiative in presenting the proposal for a course in engineering, which was set up in 1838. Chevallier had argued for it:

> 'I consider it to be also a consideration of no slight moment that the profession of Civil Engineer should take the rank in society which its importance demands, and as a preliminary step, that those who aspire to highest stations should have received the education of Christian Gentleman.'[26]

Under Chevallier's influence, the course had a strong mathematical component, but practical subjects were strongly represented, and students were also expected to attend classes in modern languages. A serious handicap to the scheme was that it did not lead to a degree. As no other university at the time offered a degree in engineering, the Durham pioneers decided that it was too great an innovation for which to take responsibility themselves, and instead they denominated the award of 'Academical Rank of Civil Engineer' for successful candidates, treating it as a degree but without calling it such. Another problem was the cost which, at around £80-£100 a year for tuition and accommodation, was about double the cost for an arts student, even though it compared favourably with the £200-£300 a year which could be charged as premiums for an apprenticeship in an engineer's office. However, the most serious weakness of the Durham scheme was that few if any professional engineers recognised it as an alternative to a 'normal' apprenticeship, so that students found themselves having to begin an apprenticeship after having spent three expensive years at the university. Numbers of students enrolling for the course were thus never high, (although in the early years they are reported as having exceeded the number for theology), and by 1851 they had declined to such an extent that no further students were registered. Well-disposed engineers who had assisted with examining the students included James Walker and Sir John Rennie, and the local mining and railway engineers Nicholas Wood and Thomas Sopwith. Sopwith was involved in a scheme to revive the course in 1858, but this soon collapsed like its predecessor, and no further engineering course was started at Durham until 1965.[27]

One university which Walker overlooked in his review of 1841 was Glasgow, which had the previous year become the first university outside London to possess a Professor of engineering when Queen Victoria instituted there a Regius Chair of Civil Engineering and Mechanics. The first incumbent of the office was Lewis D. B. Gordon, CE, but the appointment was not a success. Within days of taking up the post, Gordon complained to the Principal that he had not been provided with a room of his own but had been allocated a corner of the chemistry laboratory.[28] He enlisted the support of the Lord Advocate, but it became clear in the subsequent altercation that the university objected strongly to being expected to find additional accommodation for new external appointments at a time when resources had been curtailed, and that Gordon was a victim of in-

ternal political disagreements.[29] Despite the lack of positive evidence it seems reasonable to assume that Gordon did not receive much encouragement to devote time or attention to his academic duties. He certainly developed a flourishing practice as an
engineer during his tenure of the chair. He had been appointed at the remarkably early
age of twenty-five, having served as an Assistant Engineer to Marc Brunel on the Thames
Tunnel, and he had studied at the School of Mines at Freiberg in Saxony. He became involved in work on the Glasgow water supply and the great 447 ft chimney at the St. Rol-
lox chemical factory of Messrs Tennant, as well as on railways and on wire rope and cable
manufacture further afield. He decided to retire in order to concentrate on these business interests in 1855, and was succeeded by his deputy W. J. M. Rankine, who managed
to convert what had hitherto been a somewhat nominal appointment into a highly regarded school of engineering.[30] But this development belongs firmly to the second stage
of the transformation of the engineering profession.

The example of Glasgow was closely followed in Dublin, where Trinity College had
established a School of Engineering in 1841.[31] The initiative here was taken by Dr Hum-
phrey Lloyd, the Professor of Natural and Mechanical Philosophy, who was the main advocate for the scheme with the College authorities. John Benjamin MacNeill
(1793-1880), who was then active in Ireland building railways, was appointed the first
Professor in the School in 1842, although like Gordon in Glasgow he found it difficult to
perform this role while carrying on his professional commitments so he withdrew in 1846.
He was succeeded by Dr Samuel Downing, who was largely responsible for building up
the School into a substantial and highly regarded centre of engineering education. It had
been determined at the outset to link the course for a Licence or Diploma in Civil Engineering with a degree in the arts, and in the twenty years 1841-61 some 130 Diplomas
or Licences were awarded, the graduates going on to active professional careers at home
and abroad. In 1861, when it became possible to add a new degree course for 'Master of
Engineering', it was estimated that twenty graduates of TCD were engaged on Indian
railways and geological surveys.[32] Early graduates included James Barton (1826-1913),
who had been a pupil of MacNeill and graduated in 1848, after which he was engaged by
MacNeill on the Belfast Junction line with its novel triple-span lattice-girder Boyne Via-
duct, and the Northern Grand Junction Railway. Bindon Blood Stoney (1828-1909) was
another early product of the School, graduating with distinction in 1850 and going on to
be Assistant to the Earl of Rosse at the Parsonstown Observatory before working on
Spanish and Irish railways and becoming Chief Engineer to the Port of Dublin.[33] The
Engineering School at TCD provided a home for several years to the Institution of Civil
Engineers of Ireland, with which it had a close and productive relationship. The School
may be regarded as one of the success stories of Irish engineering, and it made a substantial contribution to the development of scientific engineering in Britain.[34]

Apart from the pioneering developments which have been described, however, all
that had been achieved by the middle of the nineteenth century towards placing British
engineering on more theoretical scientific foundations was some hesitant staking out of
the ground for possible developments. There was as yet no coherent programme for
change, nor an identifiable group of senior engineers who were prepared to promote it.
While some leading engineers were not unsympathetic to the growing need presented by

the emergence of new skills in electrical and chemical engineering for more systematic formal instruction in engineering training, and gave their support to experimental courses in Durham and elsewhere, the profession as a whole lacked any strong incentive for change. The system of training by apprenticeship had become strongly established, and the pupilage fees provided a powerful vested interest against change. The profession was, moreover, enjoying an unprecedented expansion both in numbers and in prestige, on the crest of the boom in railways, steam navigation, and public works. There seemed little reason to change what had become a demonstrably successful form of organisation and training.

John Fowler, later Sir John and designer of the Forth Railway Bridge, gave his Presidential Address to the Institution of Civil Engineers in January 1866. It was in most respects an uninspiring piece of Victorian prose, dwelling at length on a banal review of the history of the Institution and the variety of works which an engineer could be required to perform. But he also touched on the problem of the education of an engineer, expressing an earnest desire to provide:

'for the rising generation those better opportunities and that more systematic training for which in our time no provision had been made, because it was not then so imperatively required.'[35]

The requirement arose, on Fowler's diagnosis, from the need to compete with foreigners on equal terms. His suggested system, however, made no significant innovations, depending mainly on a conscientious application of school learning and practical experience through pupilage. He recognised that a period 'at one of our great universities' might be helpful , but he did not regard it as essential, and curiously Fowler envisaged a standard Oxbridge degree course for young gentlemen rather than a specialised course in engineering theory, on which he had nothing to say. The emphasis throughout his address remained on the great advantage of practical experience.

While Fowler may be regarded as having, like many of his contemporaries in the 1860s and 1870s, an anxious perception of the new demands being made on the engineering profession, requiring ever greater specialist expertise and theoretical competence, he had nothing to suggest to meet these demands other than more of the previous 'best practice'. Other engineering commentators did not do much better although some interesting ideas were canvassed. Shortly after Fowler's Presidential Address, Zerah Colburn, the vigorous editor of *Engineering*, allowed himself space in that journal to present his views on 'Engineering Education'. He began with a traditionalist statement: 'The knowledge which the youth, intended for an engineer, should acquire, would, we may believe, be best imparted by an engineer...'. But he deplored the lack of methodical instruction in the offices of most engineers:

'In the profession it is notorious that the most that can be done for the pupil is to give him opportunities for 'picking up' what he can for himself...'

To remedy which, he suggested a special school, to be established with the support of professional engineers:

'Were it given out that engineers in large practice would select their pupils mainly from those who had passed the Engineers' College, the attendance there ... would soon be such as to render any pecuniary application to the profession or the public unnecessary.'[36]

Like other good ideas, however, there was no immediate response to this suggestion from the profession.

The general background of mounting national anxiety about inadequate technical education in Britain in the second half of the nineteenth century has been well reviewed elsewhere, so that it may be taken for granted here.[37] It led to the creation of a national system of primary and technical education by the end of the century. The engineering profession took little positive part in this development, being reluctant to overcome its inhibitions about theoretical education and to encourage it openly. The Institution of Civil Engineers undertook a substantial review of 'The education and status of civil engineers, in the United Kingdom and in Foreign Countries' in 1870, the result of which was a report which retained a pervasive sense of complacency, as when it considered the existing situation in Britain:

'It is hardly necessary to remark, except for the purpose of completing the comparison with foreign countries, that in England the profession of engineering is entirely unconnected with the Government, there being no state corps of Engineers other than those attached to the Army. It is open to any one to enter the profession, and to obtain in it any standing his merits may entitle him to, and all the civic works of the country, whether public or private are (with some few exceptions, where Royal Engineers have been employed) executed by private practitioners ... There is, further, in England no public provision for engineering education. Every candidate for the profession must get his technical, like his general education, as best he can; and this necessity has led to conditions of education peculiarly and essentially practical, such being the most direct and expeditious mode of getting into the way of practical employment ... The education of an Engineer is, in fact, effected by a process analogous to that followed generally in trades, namely, by a simple course of apprenticeship, usually with a premium, to a practising Engineer; during which the pupil is supposed, by taking part in the ordinary business routine, to become gradually familiar with the practical duties of the profession, so as at last to acquire competency to perform them alone, or, at least, after some further practical experience in a subordinate capacity ... It is not the custom in England to consider *theoretical* knowledge as absolutely essential ... '[38]

This statement is worth quoting at length because it expresses so clearly the traditional view of the engineering profession in the 1870s, and demonstrates the problem facing those who wished to change the system. The scientific engineers who worked quietly and ultimately successfully to modify the deeply engrained traditional suspicion of theoretical education for engineers wisely avoided a direct confrontation with this attitude, but attempted rather to argue that, however valuable practical experience may be, it was desirable to back it up with some theoretical competence. The first of the great exponents of engineering science in the British universities was William John Macquorn Rankine. He was thirty-five when he succeeded Gordon to the Regius Chair of Civil Engineering and Mechanics at Glasgow in 1855. He had studied Natural Philosophy at Edinburgh and had served with his father in superintending the construction of part of the Edinburgh and Dalkeith Railway, and with Sir John MacNeill, the eminent Dublin civil engineer. He had also submitted prize-winning papers to the Institution of Civil Engineers and had done some brilliant innovative work on thermodynamics which had won him a Fellowship of the Royal Society in 1853. In continued association with Gordon, and with William

Thomson (later Lord Kelvin), who had become Professor of natural philosophy in 1846, and with Thomson's elder brother James, who had attended Gordon's classes and then set up as a successful professional engineer, Rankine became a member of the influential coterie which established the University of Glasgow as a world-famous centre of applied science.[39]

Rankine applied himself to the task of providing a complete range of text books for his engineering students. These included his *Manual of Applied Mechanics* (1858), *Manual of the Steam Engine and other Prime Movers* (1859), *Manual of Civil Engineering* (1862) and *Manual of Machinery and Millwork* (1869), all of which ran to many editions and remained in print well into the twentieth century as standard textbooks for all university-trained engineers.[40] He also maintained a steady output of papers in the learned journals, so that engineering theory acquired an authority which it had not possessed before. At the same time, Rankine took a keen interest in the institutional development of the engineering profession, becoming in 1857 one of the leaders in the foundation of the Institution of Engineers and Shipbuilders in Scotland, and the first President of this institution. Strangely, however, he seems to have dropped out of membership of the Institution of Civil Engineers about the same time, apparently because of a dispute about the status of his membership, but it is not being unduly imaginative to see in this incident a reflection of the traditional opposition to theoretical education amongst engineers.[41] Rankine campaigned vigorously for the recognition of engineering studies as a full university degree, and was successful in 1872, the year when he died at the early age of fifty-two.[42]

It has been argued that Rankine aimed at establishing engineering science as an intermediate mode of knowledge between pure science and pure practice, and one which would be based on the universities.[43] Rankine's Introductory Address as President to the Institution of Engineers and Shipbuilders in Scotland certainly specified a series of attainable objectives for engineering in a way which suggested a different sort of intellectual exercise from either the physical sciences or traditional engineering. After emphasising the role of economy in the scientific practice of the useful arts, Rankine went on to enumerate some of the engineering issues calling for elucidation in the future discussions of the Institution. These included the properties of materials, the means of performing useful work, electromagnetic engines and electric telegraphs, sanitary engineering, accurate workmanship and a standardised system of measurement, and legislation affecting engineering such as patent law.[44] It is not necessary, however, to speculate about Rankine's intentions. For our purposes, his important achievement was that of establishing engineering as a university study with a sound theoretical basis and appropriate recognition in a degree or an equivalent award. To the extent that it was being accepted as a systematic intellectual pursuit, engineering began to acquire in Rankine's department at Glasgow the characteristics of a branch of science. And it provided a model which was capable of detailed replication.

Nevertheless, the initial response to the Rankine system was cautious, and it was only the reform of existing universities in the second half of the nineteenth century and the establishment of new institutions of higher education that created the conditions in which engineering could take root as a generally accepted scientific discipline in Britain. Apart from Glasgow, the main focus of scientific engineering in this period was Manchester, where a sturdily independent tradition of pioneering science had developed early in the century and had been institutionalised in bodies such as the Manchester Lit. and Phil. and the Owens College, founded in 1851.[45] Despite the importance of engineering in the Manchester area, and the presence of prominent engineers such as Fairbairn, Whitworth, and Nasmyth in local affairs, Owens College did not show great interest in engineering education in its early years. But some local firms such as Beyer Peacock made generous financial grants, and Whitworth endowed the scholarships named after him which gave a boost to the reputation of engineering.[46] A new chair of engineering was established in March 1868, signalling a determination to respond to 'the development of engineering as a profession and discipline',[47] and the first appointment was Osborne Reynolds. Reynolds took some time to settle down in his office, but he endured long enough to become an institution, modelling his successful engineering courses on Rankine's work in Scotland, and emphasising 'the application of scientific principles to engineering requirements'.[48] Together with Henry Roscoe's School of Chemistry, Reynolds' School of Engineering helped to create the image of 'the University of the Busy',[49] and thus contributed to the identification of academic engineering with the highly practical and utilitarian character of applied science.

Another significant appointment in 1868 was that of Fleeming Jenkin to the chair of engineering at Edinburgh. There had been earlier experimental courses in engineering at Edinburgh,[50] but substantial development had to await the success of Rankine at Glasgow, and Jenkin represented the Rankine system. In his inaugural address, after giving an urbane review of continental and British practice in engineering education, Jenkin came down in favour of the British method with its traditional pupilage/practical structure, only emphasising the necessity of a period of theoretical training at a university as a precondition to the success of this method. He thus avoided a head-on confrontation with the British engineering tradition while identifying its major deficiency: 'Our defect is the want of a good knowledge of the theories affecting our practice'.[51] He criticised by vivid caricature the abuses of apprenticeship, but he was firm in his support for the traditional method of instruction, arguing that all it required to make it successful was a sound theoretical preparation. This gave him and the handful of like-minded colleagues in other universities all the leverage they needed to establish strong university courses and thus to lay the foundations for the transformation, by the end of the century, of the attitudes of the engineering profession towards theoretical instruction and to place beyond dispute the scientific quality of engineering.

When Rankine died in 1872, a firm basis of engineering theory had been established in Britain and it had been recognised by a few universities as a scientific discipline. Much, however, remained to be done before the engineering profession as a whole could be persuaded to accept the value of scientific theory as an integral component in engineering instruction. The fact that this objective was substantially achieved by the end of the nineteenth century was due in part to the increasing public pressure derived from apprehension about foreign competition and manifesting itself in the devotion of public resources to technical instruction and to the higher education of engineers. But it was due even more to the success of a small elite of engineering academics in seizing the opportunities provided by this increased scope for engineering education and, in particular, to the incorporation of laboratory practice and research as a counter-part to theoretical instruction in the developing schools of engineering. The early exponents of university engineering had been starved of such basic resources as accommodation: Gordon's struggle for space in an existing chemistry laboratory in the 1840s had been indicative of a general problem, and even the successful university engineers such as Rankine and Osborne Reynolds spent most of their teaching time lecturing in small class rooms. To become completely successful, university engineering needed a build-up of staff so that schools of engineering could provide instruction throughout the undergraduate careers of their recruits, and it needed space so that the same schools could establish extensive laboratories. The resources for both became available in the last third of the nineteenth century, and a number of talented engineers were in positions where they were able to utilise these resources.

The London colleges had been overtaken by provincial developments in the middle of the nineteenth century, but in the last third of the century they reasserted their leadership in engineering education. At University College, the chair of civil engineering was held from 1859 to 1867 by William Pole, who was also Honorary Secretary to the Institution of Civil Engineers and the biographer of Sir William Fairbairn. A new chair of applied mathematics and mechanics was established in 1868, and A. B. W. Kennedy came to hold this together with the civil engineering chair in 1874. In 1878, Kennedy held the new chair of mechanical technology as well, until he resigned in 1889 to devote himself to private practice while he was still a young man of forty-two. Kennedy's great achievement at UCL was to establish the first genuine engineering teaching laboratories. Other university engineering departments had already begun to experiment with the use of laboratories, but it was Kennedy who made them an indispensable part of instruction in scientific engineering. The steady growth and success of engineering studies at UCL was recognised by the setting up of a separate Faculty of Engineering in 1908.[52]

Having fluctuated around fifty in the middle decades of the century, the numbers of engineering students at King's College London began to rise again in the 1880s, from sixty-nine in 1881 to over a hundred in 1883. New workshops had been opened in 1869, and in 1890 a substantial laboratory was acquired for electrical engineering. Engineering became a separate Faculty at the College in 1896, (twelve years before UCL), recognising the established importance of the discipline which in 1893 had the unusual honour of being able to claim that the Presidents of the three leading institutions were all old KCL students: Harrison Hayter of the Civils, William Anderson of the Mechanicals, and W. H. Preece of the Electrical Engineers.[53] Under John Hopkinson, Professor of electrical engineering in 1890, the college possessed the largest school of engineering in Britain.[54]

Amongst other London colleges to encourage the growth of scientific engineering in these years, the Central Institution of City and Guilds Institute opened in South Kensington in 1884, becoming the Central Technical College in 1893 and a constituent part of Imperial College in 1910. At its foundation, William Crawthorne Unwin was appointed Professor of civil and mechanical engineering and thus Head of the principal department of the college. Under Unwin's leadership, it won an impressive reputation for scientific engineering, and he was appointed as the first University of London Professor of engineering in 1900, although he retired from teaching four years later.[55] The City and Guilds also established evening classes in engineering at Finsbury in 1878 which grew into the Finsbury Technical College in 1883. The staff included W. E. Ayrton and John Perry, who had both served under Henry Dyer in the Tokyo Engineers' College when this had opened in 1873. Perry became Professor of mechanical engineering at Finsbury and put into practice Dyer's principles of heuristic instruction and sandwich courses.[56]

While London thus gave a strong lead in the development of fully integrated faculties of academic engineering in the last third of the nineteenth century, new institutions of higher education were taking root in the larger industrial cities, all of which had a predisposition in favour of engineering as a practical, vocationally orientated subject, congenial to the successful industrialists and businessmen who supplied much of the original endowments. The pioneering work of Osborne Reynolds at Owens College, Manchester, has already been mentioned. His example was closely followed in Leeds, Liverpool, Sheffield, Birmingham, Nottingham, Newcastle, Bristol and Southampton, all of which achieved well developed schools of engineering by the First World War.[57] There was some local specialisation: a chair of naval architecture at Liverpool;[58] research in gas engineering at Leeds;[59] a chair of mining at Newcastle;[60] and some early work on electrical engineering at the Hartley Institution in Southampton.[61] But the common basis of all these initiatives was training in engineering theory and practice with the help of well-equipped laboratories. The laboratories at Liverpool were built with money from local brewers and shipowners, first in 1889 and then extended in 1910.[62] Firth College, Sheffield, opened in 1880, with an engineering laboratory provided at the outset.[63] In Bristol, the Merchant Venturers' Technical College provided the Faculty of Engineering in the new university after 1909, as a result of large endowments by the Wills Family.[64] Edinburgh University, where Fleeming Jenkin established a sound foundation of instruction in engineering theory between 1868 and 1885, did not possess an engineering laboratory until 1890.[65]

Oxford was slow to respond to the growing academic interest in scientific engineering. Its first Professor of engineering was appointed in 1908. He was C. F. Jenkin, the son of Fleeming Jenkin.[66] At Cambridge, the response had been earlier and more generous. Some engineering had been taught at the university before the foundation of the first chair in mechanism and applied mechanics in 1875, and it had been an examinable subject from 1865. The first holder of the chair was James Stuart, who did much to encourage the development of higher education elsewhere through his programme of Extension Lectures. He was succeeded in 1890 by Alfred Ewing, a brilliant and articulate Scots engineer who moved from Dundee to take the Cambridge chair and held it until 1903.[67] A laboratory was opened at Cambridge in 1894, and the Engineering Tripos began about the same time, thus fulfilling the main points in the programme set out by Ewing in his inaugural lecture of January 1891. This had been a clear and functional address recapitulating the case, which Ewing regarded as proven, for theoretical university education to precede and shorten pupilage, but not to replace it. He had described the rising influence of schools of engineering elsewhere, and especially the value of laboratories, following Kennedy of whom Ewing said: 'He found the engineering laboratory no more than a means of private research. He left it an instrument of education'. Ewing placed great emphasis on the role of laboratories:

> 'It is scarcely too much to say that the engineering laboratory has done for the scientific teaching of engineering what the physical laboratory has done for the teaching of physics or the chemical laboratory for the teaching of chemistry ... The engineering laboratory has in fact gone far to bridge whatever gulf may be said to have existed between professors and practitioners.'

This judgement led him on to appeal for resources to undertake at Cambridge the developments which his energetic prosecution then carried through. He concluded:

> 'Applied mechanics does not now knock at the doors of the universities, supplicating admission. The question is not whether a school of engineering is a legitimate part of a university, but whether a university is reasonably complete without a school of engineering.'[68]

The case for an academic and theoretical component to engineering in Britain was thus effectively made and triumphantly vindicated in the second half of the nineteenth century, and this success transformed the engineering profession. What had still been essentially a traditional craft based upon the acquisition of skill by practice as late as the Great Exhibition of 1851, was converted by the outbreak of the First World War in 1914 into a sound theoretical discipline based on scientific principles, albeit in a highly practical context. The crucial change for the profession was the recognition of theoretical instruction tested by examination as a condition for admission to the leading institutions. The Institution of Municipal Engineers had pioneered this innovation with the introduction of its 'Testamur' examination in 1886,[69] but this case was unusual to the extent that the backgrounds of would-be municipal engineers were more diverse than occupations requiring specific technical skills. However, the precedent was followed by other professional groups, especially by the Institution of Civil Engineers, which accepted the desirability of university qualifications in engineering for admission to membership in 1897. This did not automatically supersede traditional pupilage, with all its prejudices and vested interests, but it was a decisive step towards the full acceptance of theoretical qualifications gained through an academic process. These were incorporated in the standards laid down for admission to the ICE by the Special Committee on the Practical Training of Engineers in 1909.[70] Professor Unwin had vigorously promoted the benefits of college training in these discussions. His views were that:

> '... it is more and more recognised that although an engineer cannot be made in college, yet a college education is an essential part of the training of an engineer.'[71]

The procedure adopted by the Civils was then gradually adopted by the other professional institutions of engineers, so that the desirability of academically-based principles of scientific engineering became generally accepted.

By 1914, therefore, a remarkable transformation had been achieved in British engineering education. The establishment of chairs in engineering had not immediately guaranteed the necessary facilities for instruction, nor an intake of students to benefit from them, and there had often been a long time-lag between the appointment of a professor and the provision of adequate laboratory accommodation. Such provision did not become a matter of course until the end of the nineteenth century, by which time funds had been made available for engineering laboratories, both through the provisions of the Technical Instruction Act of 1889 and through the initiative of private endowments. By this time, also, the number of students had increased substantially. The pioneering Durham course in the 1840s had produced only two or three examinees in engineering each year in the decade.[72] By the end of our period, however, the engineering departments at KCL and UCL had been recognised as independent faculties, and there were well-established engineering schools at Glasgow, Manchester, and Cambridge universities. It has been estimated that the overall total of engineering students at any one time in the years immediately before the First World War was about 1,500.[73]

Credit for this achievement belongs largely to the small groups of men who staffed the university departments of engineering in these years, and especially to those who held chairs of engineering. They were not carrying out a well-conceived programme in which every stage was carefully planned and fulfilled. Rather, they were responding in a pragmatic way to a growing recognition of the theoretical inadequacy of traditional forms of engineering instruction. The first generation of university engineers, inspired by Rankine, struggled against the predominantly craft bias of a method of training based entirely on pupilage to establish a few centres of theoretical instruction for British engineers. Their successors such as Kennedy, Ewing and Unwin built on these foundations to construct a comprehensive system of engineering education which became a prerequisite for virtually everyone wanting admission to the profession. The ability of these men to grasp the opportunities offered them brought about a quiet revolution in the British engineering profession and made certain that engineering instruction would in future be firmly based on scientific principles and theoretical competence. In conclusion, however, it is important to emphasise that this very considerable achievement of the university engineers was essentially a compromise. It involved a recognition of the remarkable strength and resilience of the British engineering tradition of instruction through pupilage, which was never discarded but accommodated to the acknowledged need for more theoretical knowledge by having a university training or its equivalent grafted on to the existing apparatus of apprenticeship and collective self-education through membership of the professional institutions. Thus by 1914, thanks largely to the vision and diplomacy of men like Fleeming Jenkin and Alfred Ewing, the British engineers had the opportunity to enjoy the best of both worlds - the confidence of theoretical expertise combined with sound practical experience. This compromise was a distinctively British solution to the problems of education and training, and one which should have enabled British engineers to face the challenges of the twentieth century with confidence. The fact that they did not do so cannot be attributed to a failure to adapt to the educational requirements of the new century.

Notes

1. For both Kernot and Higinbotham, see *Australian Dictionary of Biography*: the anecdote appears in the entry for Kernot.

2. The best general survey is D. S. L. Cardwell, *The organisation of science in England*, 1957. See also E. Ashby, *Technology and the Academics*, 1958; M. Argles, *South Kensington to Robbins*, 1964; and M. Sanderson, *The Universities and British Industry 1850-1970*, 1972. W. H. G. Armytage, *A Social History of Engineering*, 1961, reviews the development of engineering in a broad social context, with perceptive observations on educational aspects but no detailed treatment.

3. See above, Chapters Four and Five.

4. Whitworth Porter, *History of the Corps of Royal Engineers*, 1889, 2 vols: 2, Chapter IV, 'the School of Military Engineering'. The college was established in 1812 (p. 72). The first Director was (Sir) Charles Pasley, RE, who subsequently became involved in civil engineering as an inspector of railways. Several other military engineers made this transition, but they remained a very small minority in the growing profession, and most of the early inspectors were trained at the Royal Military Academy, Woolwich, which had been founded in 1741. Very few British engineers were trained overseas, apart from immigrants such as (Sir) Marc Brunel, who sent his son I. K. Brunel to France to complete his engineering education: see R. A. Buchanan, 'Science and Engineering: a case study in British experience in the mid-nineteenth century', in *Notes and Records of the Royal Society of London*, 32, no. 2, March 1978, 215-223. Lewis Gordon, CE, the first Professor of engineering at Glasgow University, was trained at Freiberg, Saxony, but few of his contemporaries had similar experience.

5. A. W. Skempton (ed.), *John Smeaton FRS*, 1981: especially Chapter II by Norman Smith, 'Scientific Work' (35-57).

6. R. E. Schofield, *The Lunar Society of Birmingham*, 1963.

7. A. E. Musson and Eric Robinson, *Science and Technology in the Industrial Revolution*, Manchester, 1969: Chapters 8, 9 and 10 deal particularly with this theme.

8. W. Smith made many important geological observations while working as engineer to the Somerset Coal Canal and as a surveyor in the North Somerset Coalfield. See Joan M. Eyles, 'William Smith: some aspects of his life and work', in Cecil J. Schneer (ed.), *Toward a history of geology*, Cambridge, Mass., 1969, 142-158.

9. A. W. Skempton, 'Telford and the designs for a new London Bridge', Chapter 4 in Alistair Penfold (ed.), *Thomas Telford: Engineer*, 1980, 76-8.

10. The relationship is expounded in D. S. L. Cardwell, *From Watt to Clausius: the rise of thermodynamics in the early industrial age*, 1971.

11. See R. A. Buchanan, 'Science and Engineering ...', note 4 above.

12. J. Morrell and A. Thackray, *Gentlemen of Science: early years of the British Association for the Advancement of Science*, Oxford, 1981, 473-4.

13. T. M. Charlton, 'Theoretical Work', chapter IX in Sir Alfred Pugsley (ed.), *The Works of Isambard Kingdom Brunel*, 1976, 183-202.

14. I. K. Brunel, *Private Letter Books* VI 2 December 1848. The *PLB* are in the archives of Bristol University Library.

15. I. Brunel, *Life*, p. 192: the letter is also given in full, 486-489. For further comment on this discussion, see chapter ten below.

16. There is a detailed account of these experiments in Nathan Rosenberg and Walter G. Vincenti, *The Britannia Bridge: the generation and diffusion of technological knowledge*, Cambridge, Mass., 1978. See also William Pole (ed.), *The Life of Sir William Fairbairn, Bart.*, 1877, reprinted Newton Abbot 1970, p. 201 etc.

17. George S. Emmerson, *John Scott Russell: a great Victorian engineer and naval architect*, 1977, 12-24.

18. George S. Emmerson, *John Scott Russell*, p. 182, refers to this work as 'a *Great Eastern* of books'. Russell also became an enthusiastic advocate for the creation of the Royal School of Naval Architecture.

19. There is no full biographical treatment of William Froude, but see the *DNB* article for an outline of his career. See also *The Papers of William Froude MA LLD FRS 1810-1879*, Inst. of Naval Architects, London, 1955.

20. *Proc. Institution of Civil Engineers*, 1841, 25-6.

21. K. H. Vignoles, *Charles Blacker Vignoles, Romantic Engineer*, Cambridge 1982, 98-100.

22. H. H. Bellot, *University College London 1826-1926*, 1929.

23. F. J. C. Hearnshaw, *The centenary history of King's College London, 1828-1928*, 1929, p. 147.

24. W. H. G. Armytage, *A Social History of Engineering*, 1961, 150-1.

25. A. L. Turner (ed.), *History of the University of Edinburgh 1883-1933*, Edinburgh 1933. See also: T. Hudson Beare, *The Education of an Engineer*, Edinburgh 1901.

26. C. Preece, 'The Durham Engineer Students of 1838', in *Transactions of the Architectural and Archaeological Society of Durham and Northumberland*, new series 6, 1982, 71-74. The 'Chevalier Correspondence' is in the University of Durham Library.

27. But its place was largely taken in North East England by the College of Physical Science founded at Newcastle in 1871. This was the result of close collaboration with the North of England Institute of Mining and Mechanical Engineers, which even shared the same registrar for some years: see E. M. Bettenson, *The University of Newcastle upon Tyne*, Newcastle 1971.

28. Glasgow University Archives: copy letter P/CN/Macfarlan 469/470.

29. This is indicated by correspondence between the Principal and the Lord Advocate: see Glasgow University Archives, P/CN/MacFarlan 474 (8th December 1840) and 479.

30. James Small, 'Glasgow University's contribution to engineering progress', in *Glasgow University Engineering Society Yearbook*, 1954, p. 27.

31. For a general account of the development of engineering in Ireland, see above, Chapter 7, especially notes 1-12. Also Noel J. Hughes, *Irish Engineering 1760-1960*, Institution of Engineers of Ireland, 1982; and Ronald C. Cox (compiler), *Engineering Ireland 1778-1878*, Exhibition Catalogue, Trinity College Dublin, 1978.

32. See Sir Richard Griffiths, Presidential Address, *Trans. Inst. of Civil Engineers of Ireland*, 6, 1859-61, 204.

33. See Cox, *op. cit.* 52-3.

34. For a summary of the educational developments at TCD, see John Purser, 'A Note on the Engineering School since its Foundation' in *Hermathena*, 58, November 1941, 53-56. giving a list of the Professors of Civil Engineering ending with himself.

35. *Proc. Institution of Civil Engineers*, 25, 1865-6, 203-228.

36. *Engineering*, 2 Feb. 1866, p. 79.

37. This national anxiety is examined by E. Ashby, *Technology and the Academics*, 1959: see his Appendix giving Lyon Playfair's letter of 7 June 1867 to Lord Taunton regarding the poor performance of Great Britain at the Paris Exhibition of 1867 (pp, 111-3). See George S. Emmerson, *Engineering Education - a Social History*, Newton Abbot, 1973.

38. *The education and status of civil engineers, in the United Kingdom and in Foreign Countries*: Report by the Institution of Civil Engineers, 1870, viii-ix.

39. See James Small, 'Glasgow University's contribution ...', note 30 above.

40. David F. Channell, 'The harmony of theory and practice: the engineering science of W. J. M. Rankine' in *Technology and Culture*, 23, 1, January 1982, 39-52.

41. Hugh B. Sutherland, *Rankine - His Life and Times*, Rankine Centenary Lecture, Institution of Civil Engineers, 1972, p. 11.

42. Glasgow University Archives CIVA 36407 - Report of 29 April 1862, and Documents GS - 4600, give details of petitions and correspondence regarding a 'Certificate of Proficiency' in Engineering and the request for a Degree in the subject. See also J. D. Mackie, *The University of Glasgow 1451-1951*, Glasgow, 1954.

43. David F. Channell, 'The harmony of theory and practice ...', note 40 above. See also the same author's essay, 'Rankine, Aristotle and Potential Energy' in *The Philosophical Journal*, Transactions of the Royal Philosophical Society of Glasgow, 14, 2, 1978, 111-114.

44. *Transactions of the Institution of Engineers in Scotland*, 1, 1857, 3 et seq., 1857.

45. See R. H. Kargon, *Science in Victorian Manchester: Enterprise and Expertise*, Manchester 1977, especially Chapter 5. Also J. Thompson, *The Owens College, its foundation and growth*, Manchester 1886.

46. See F. C. Lea, *Sir Joseph Whitworth - a pioneer of mechanical engineering*, 1946. Whitworth scholarships were established in 1868 with an annual grant of £3,000, initially intended to provide thirty £100 scholarships a year.

47. R. H. Kargon, *Science in Victorian Manchester*, p. 184.

48. Ibid, p. 188. Simultaneous developments occurred in the Physics Department at Owens College, which promoted a course in electrical engineering.

49. The term was used in the *Spectator*, as quoted by M. Neve in TLS 21 July 1978, p. 879.

50. George Wilson had held a chair of technology at Edinburgh from 1855 to 1859, when it had been abolished: see A. L. Turner (ed), *History of the University of Edinburgh 1883-1933*, Edinburgh, p. 272.

51. Fleeming Jenkin, *A lecture on the education of civil and mechanical engineers in Great Britain and abroad*, Edinburgh, 1868, p. 16.

52. H. H. Bellot, *University College London 1826-1926*, 1929; and R. E. D. Bishop, 'Alexander Kennedy - the elegant innovator' in *Transactions of the Newcomen Society*, 47, 1974-76, 1-8. Kennedy was born in Stepney in 1847 and died in 1928.

53. F. J. C. Hearnshaw, *Centenary History of King's College London*, 1929; G. Huelin, *King's College London 1828-1978*, 1978.

54. M. Sanderson, *The Universities and British Industry 1850-1970*, 1972, p. 109.

55. E. G. Walker, *The Life and Work of William Crawthorne Unwin*, 1947.

56. The Japanese connection provides an interesting illustration of the international status of British engineers in the nineteenth century: see W. H. G. Armytage, *Social History ...*, p. 233; and W. H. Brock, 'The Japanese Connexion: engineering in Tokyo, London, and Glasgow at the end of the nineteenth century', in *British Journal for the History of Science*, 14, 3, no. 48. November 1981, 227-243.

57. See W. H. G. Armytage, *Civic Universities*, 1955: also P. H. J. H. Gosden and A. J. Taylor, *Studies in the history of a university 1874-1974*, Leeds, 1975; S. Dumbell, *The University of Liverpool 1903-53*, Liverpool, 1953; A. W. Chapman, *The Story of a Modern University*, Oxford, 1955; E. W. Vincent and P. Hinton, *The University of Birmingham: its history and significance*, Birmingham, 1947; A. C. Wood, *The History of University College Nottingham*, Oxford, 1953; E. M. Bettenson, *The University of Newcastle upon Tyne*, Newcastle, 1971; R. A. Buchanan, 'From Trade School to University', in Gerald Walters (ed.), *A Technological University - an experiment in Bath*, Bath, 1966, 12-26; and A. T. Patterson, *The University of Southampton*, Southampton, 1962.

58. S. Dumbell, *The University of Liverpool 1903-53*, Liverpool, 1953; the chair in naval architecture was founded in 1903.

59. Supported by the Institution of Gas Engineers. See A. N. Shimmin, *The University of Leeds*, Cambridge, 1954.

60. This was the first English chair in mining: the first incumbent was J. H. Merivale. See C. E. Whiting, *The University of Durham*, 1932, p. 197.

61. A. T. Patterson, *The University of Southampton*, Southampton, 1962.

62. M. Sanderson, *The Universities and British Industry 1850-1970*, 1972, p. 64.

63. Courses were offered in mechanics, mining, and metallurgy: see A. W. Chapman, *The Story of a Modern University*, Oxford, 1955, p. 40.

64. R. A. Buchanan, 'From Trade School to University', in Gerald Walters (ed.) *A Technological University - an experiment in Bath*, Bath, 1966, 12-26; see also B. Cottle and J. W. Sherborne, *The Life of a University*, Bristol, 1967.

65. A. L. Turner (ed.), *History of the University of Edinburgh 1883-1933*, Edinburgh, 1933.

66. Sanderson, *The Universities ...*, p. 39.

67. Ibid., p. 45. see also T. J. N. Hilken, *Engineering at Cambridge University 1783-1965*, Cambridge, 1967.

68. J. A. Ewing, *The University Training of Engineers*, Cambridge, 1891.

69. Literally, a certificate that one has passed an examination. For the introduction of the 'Testamur' examination, see P. Parr, Presidential Address, 1954, *Proc. Institution of Municipal Engineers*, 81, 1954-5, 5. See above, Chapter 5.

70. *Proc. Institution of Civil Engineers*, 178: Report of Special Committee on practical Training, 228-238 and an Appendix 247-253. See also 186, 406-421 for Report on Conference on Education and Training of Engineers, 1911.

71. E. G. Walker, *The Life and Work of William Crawthorne Unwin*, 1947, p. 147: from a speech at a distribution of prizes at the Merchant Venturers Technical College, Bristol, in 1912.

72. C. E. Whiting, *The University of Durham*, 1932, p. 81.

73. Estimates of numbers of engineers undergoing training in this period vary according to the basis of selection. Argles, *South Kensington to Robbins* ..., reckons that there were 1,487 full-time students of engineering and technology in England and Wales in 1912-13; Sanderson, *The Universities* ..., gives a figure of 1,433 students of British universities (including a substantial Scottish contribution) in 1892. Figures for numbers of students in 1913 are provided in *Reports from University Colleges 1913-14* as reproduced in M. Sanderson (ed.), *The Universities in the nineteenth century*, 1975, 243-4.

Chapter Ten

Ideas and Beliefs of the Engineers

Sir Daniel Gooch, Bt., MP (1816-1889) began his highly successful engineering career by serving under I. K. Brunel as Locomotive Superintendent to the Great Western Railway. His locomotives with large-diameter driving wheels performed outstandingly well on the broad gauge system of the GWR. Gooch followed Brunel as engineer to the steam ship *Great Eastern* and used it to lay the first successful transatlantic telegraph cable. In 1865 he became chairman of the GWR, helping to steer the company through a period of great financial difficulty, and in the same year he was elected Member of Parliament for Cricklade, the constituency which at that date included the railway town of Swindon, with its concentration of GWR engineering workshops. Gooch served for twenty years in Parliament, and on his retirement in 1885 he wrote in his diary:

> 'The House of Commons has been a pleasant club. I have taken no part in any of the debates, and have been a silent Member. It would be a great advantage to business if there were a greater number who followed my example.'[1]

This is a very personal and particular statement of opinion, but coming as it does from one of the leading engineers of Victorian Britain, and with a deficiency of more easily quantifiable evidence, it may be taken as indicative of engineering attitudes in this period. We are obliged, in fact, to rely largely on such anecdotal evidence in any reconstruction of the ideas and beliefs of the nineteenth century engineers, because they were reluctant to commit themselves, as a group or as individuals, on any contentious matter involving a non-professional judgment, such as a political or religious issue. Their professional institutions abhorred the discussion of such questions, and as few engineers expressed themselves in a literary form which has survived, any analysis of their ideas and beliefs is bound to be tentative, and to depend heavily on such fragments of evidence as are available. Nevertheless, it is possible to make some useful generalisations, and that is the purpose of this chapter. It will be suggested, in the first place, that the dominant attitude of engineers on non-professional matters was one of social conformism, and this will then be demonstrated from the political, religious, and intellectual commitments of engineers in the nineteenth century.

The apparent silence of most engineers who come within the scope of this survey on their personal ideas and beliefs suggests that such matters were not central preoccupations to them. An argument from negative evidence can rarely be conclusive, but it does seem likely that if the engineers, as individuals or as groups, had felt strongly on any of the great political, religious, or intellectual issues of the period, then some token of this commitment would have been registered for posterity. As it is, however, there is remarkably little evidence of any sort of engineering identification with these matters. A groundswell of conventional wisdom arises from the professional press, from presidential addresses to the engineering institutions, and from such personal comments as survive in memoirs, diaries, and letters, but there is scarcely anything which can be regarded as robustly forthright on any matters other than the narrow range of those such as bridge design and technical education which possessed a high professional content. This whisper of engineering opinion serves at least to demonstrate that the profession was not comatose on such matters, but it also suggests that engineers preferred, other things being equal, to get on with their jobs and to leave general social issues to others. This essential conservatism emerges as a persistent theme in the attitudes, ideas, and beliefs of British engineers in the period under discussion.

The engineers represented a particularly strong tradition of conformism in nineteenth century politics. It is virtually impossible to identify any engineer as a political radical in this period. Radicalism, of course, takes many different forms and it is not to be interpreted here as a party label: it refers instead to a readiness to express dissenting or unconventional political views. The case of Thomas Telford is interesting in this respect, because he could be regarded as a sort of radical *manqué*. As the orphaned son of a Dumfries-shire shepherd who found himself struggling against almost overwhelming social disadvantages, he had been a fervent supporter of the French Revolution in his youth. He was not the only young man of the period to be enthusiastic about this dramatically radical event:

'Bliss was it that dawn to be alive
But to be young was very heaven!'[2]

But like Wordsworth, Telford quickly outgrew the political enthusiasms of his youth, and for most of his life successfully hid any residual radicalism under a mask of political conformism.

A generation later, I. K. Brunel campaigned for his brother-in-law Benjamin Hawes in the first election for the reformed House of Commons in 1832: Hawes stood successfully for Lambeth in the Radical cause, so it is a reasonable inference that Brunel at that time was sympathetic to radicalism. Subsequently, he showed a keen interest in the French political disturbances of 1848, when a revolution in Paris inaugurated the Second Republic.[3] But such incidents were essentially diversions in a life which was committed to engineering, and they were never permitted to interfere with professional business. Brunel persistently resisted attempts to persuade him to stand for Parliament. On such political issues as he did express himself forcibly - and Brunel could certainly do that very effectively when he felt moved to do so - they were invariably matters which touched on professional interests such as the 'interference' of government in the control of bridge design and the undesirability of patent legislation, on both of which his views were very emphatic although by no means unconventional.[4] His views, indeed, were those of any main-stream Free Trade, *laissez-faire*, advocate of the period, so that they scarcely constituted an expression of political radicalism at a time when such views were common currency amongst a majority of the influential figures of British public life.

If Telford and Brunel cannot convincingly be cast as political radicals, there is little chance of finding other prominent engineers who are likely to fill the role, as most of their contemporaries were politically more cautious, more conformist, and more conservative than they were. Telford's fellow-Scotsmen James Watt and John Rennie expressed no political opinion other than a pervasive conservatism. John Smeaton and James Brindley are not remembered for any political views. George Stephenson had no time for politics. Robert Stephenson did consent to becoming MP for Whitby, which he represented in the Conservative interest, and he did on occasion speak quite forcibly in Parliament, usually on issues involving engineering expertise such as the plan to construct a canal across the Isthmus of Suez, which he opposed.[5] Joseph Locke also became an MP, sitting as a Liberal for Honiton in Devon, where he had bought the Manor. But he was not a good party man and when he chose to speak in the House of Commons it was usually on matters of railway finance and accounting, on which he did not win much popularity.[6]

Engineering political commitments became, if anything, even less conspicuous after the days of the railway pioneers. John Fowler, engineer of the Metropolitan Railway in London, stood as a Conservative MP, but was not elected.[7] And even though he succeeded in becoming an MP, Daniel Gooch, as we have observed, preferred to keep his political counsel to himself. At least Gooch appreciated the House of Commons as a convenient club, but in the intensely partisan atmosphere of twentieth century politics the office of MP has come to involve more administrative drudgery and committee work than would be congenial to most engineers. Under these conditions it is virtually impossible for an engineer in Parliament to maintain his professional practice in the way that lawyers are still able to, and this has been a further disincentive against political involvements. It is scarcely surprising, therefore, that there are only eight Chartered Engineers in the House of Commons in the 1980s, however much spokesmen of the profession might deplore the fact.[8]

The attitudes of nineteenth century engineers towards local government and the magistracy tended to reflect this general disinclination for personal involvement. William Jessop was mayor of Newark in 1790-1 and 1803-4, and he served briefly as a justice of the peace in the town in 1805.[9] Those engineers who aspired successfully to the trappings of gentility occasionally accepted the duties as magistrates which accompanied this social status. Sir Francis Fox referred incidentally to having served 'as a magistrate of forty-nine years' standing' in the course of reviewing his long career in engineering,[10] but the paucity of such references in the literature make it appear that such cases were the exception rather than the rule. In general, it is reasonable to conclude that engineers have acquired a marked antipathy towards any sort of public service unconnected with their professional duties.

As far as political opinions were concerned, most engineers of the period seem to have shared the prevailing assumptions and prejudices of their contemporaries. On matters of imperialism, for instance, engineers who went to serve in the colonies during the nineteenth century usually managed to adopt the chauvinist consensus without difficulty. John Brunton wrote about his experiences in India with the assumption of British superiority and contempt for native traditions which was characteristic of the Raj, and his attitude may be regarded as typical.[11] Although Brunton and a host of other engineers, both civil and military, did invaluable service in providing transport, irrigation, and urban amenities, they never expressed any radical reservations about the imposition of British power on India. In the more confused situation in China in the wake of the Taiping Rebellion, the British military engineer Charles George Gordon won a distinguished reputation for himself as a restorer of law and order and government authority, and acquired the nickname 'Chinese Gordon' before his fate at Khartoum in 1885 ensured his fame in imperial history for other reasons.[12]

Imperialist assumptions amongst nineteenth century engineers were buttressed by a strong sense of militarism. By the time he became President of the Civils in 1869, Charles Blacker Vignoles held one of the oldest commissions in the British army.[13] This was because he had been commissioned as an ensign when he was still an infant, to assist his repatriation when his parents died of yellow fever in the West Indies. He was old enough to have seen active service at the end of the Napoleonic Wars, and although relegated to half-pay in 1817 he retained the rank of Captain and was not averse to using his military uniform to assure potential clients such as the autocratic Tsar Nicholas I that he was politically reliable.[14] Charles Manby, Secretary to the Civils, acquired the military rank of Colonel for his exertions in the formation of an Engineer and Railway Volunteer Staff Corps.[15] Other leading engineers joined this body, including John Fowler who joined in 1865 and continued to take an interest in it for the rest of his life, becoming its Commandant in 1891.[16] The Corps enjoyed a considerable following in the railway towns such as Crewe.[17] The engineers thus demonstrated their readiness to do their duty for King and Country long before the great patriotic call of 1914 summoned so many of them to their deaths.

In all these expressions of political assumptions there was a pervading sense of ideological conformism. No hint of revolutionary, Marxist, or even sentimental socialist thinking is apparent in the surviving literature. It is true that William Morris, the pioneer British socialist, makes a brief appearance in the biography of the highly traditionalist John Fowler, but this is only as a critic of the aesthetic qualities of the Forth Bridge.[18] Generally speaking, however, and it is with generalisations that we are concerned here, engineers have shown themselves to be excellent 'organisation men', subordinating their personal views to the need to maintain a consensus amongst a diverse group of people in pursuit of a specific practical objective. Their need to serve clients who are agents of government or large and responsible bodies, and their keen appreciation of material and practical problems, has accommodated them comfortably to the management of large organisations. They had neither the time nor the inclination to cultivate radical objectives in the nineteenth century, but they took in their stride the developments which transformed government and business organisation in this period, and by making themselves available to new agencies of government some engineers found a way of becoming important public servants.

The most significant achievement of the engineering profession in this respect was in the inspectorates and public commissions which became an increasingly prominent feature of public life in the second half of the nineteenth century. The first and most important of these functions was the staffing of the Railway Inspectorate. This had been created by the Railway Regulation Act of 1840 which had provided for the appointment of an Inspector General of Railways with the duty of approving new lines for the Board of Trade. A subsequent Act in 1844 added the duty of certifying rolling stock for passenger transport. The Inspectorate was part of the Board of Trade, except for the years 1846-51 when it operated under the Commissioners of Railways, after which it was re-integrated into the Board as the 'Railway Department'. The 1840 Act required that persons appointed as inspectors should not be connected with a railway company, and even though this provision was soon withdrawn the practice of making appointments only from the ranks of the Royal Engineers was maintained from the outset. This became, in effect, the major contribution of military engineering to civil practice in Britain, although it should be observed that most of the railway inspectors became members of the Civils and that few of them returned to military service. The initial appointments were distinguished senior officers: first, General Sir John Mark Frederic Smith (1790-1874), who served for the year 1840-41 and then left to become Director of the Royal Engineers' Establishment at Chatham; and second, Major General Sir Charles W. Pasley (1780-1861), who had held the post at Chatham before Smith, and who succeeded him as Inspector-General of Railways for the years 1841-46.[19] An Assistant Inspector was appointed in 1844, of a less elevated rank: he was Captain Joshua W. Coddington (1802-53), who served until 1847. A third post was created in 1853, and thereafter the standard practice for the remainder of the nineteenth century was to have three inspectors in office, usually appointed at the rank of Captain although often subsequently promoted. Some, like Captain John L. A. Simmons (1821-1903), went on to distinguished careers with the Royal Engineers elsewhere. Simmons became an Associate Member of the Civils and took part in the animated discussion at the Institution in 1850 which arose from his refusal to approve the Torksey girder bridge over the Trent for public traffic.[20]

The question of railway bridges had become a sensitive one after the collapse of Robert Stephenson's bridge over the River Dee at Chester on 24 May 1847 with the loss of five lives. Public concern about the accident was so great that the government took the novel step of appointing a Royal Commission 'to inquire into the Application of Iron to Railway Structures'.[21] The Commission was not generally welcomed by the engineering profession, and I. K. Brunel delivered a particularly vehement attack on the notion that a government-sponsored enquiry could do anything useful in the circumstances:

> 'If the Commission is to enquire into the conditions "*to be observed*", it is to be presumed that they will give the result of their enquiries; or, in other words, that they will lay down, or at least suggest, "rules" and "conditions to be (hereafter) observed" in the construction of bridges, or, in other words, embarrass and shackle the progress of improvement tomorrow by recording and registering as law the prejudices or errors of today... Devoted as I am to my profession, I see with fear and regret that this tendency to legislate and to rule, which is the fashion of the day, is flowing in our direction.'[22]

Brunel's views represented the dominant *laissez-faire* consensus of his generation, and anticipated the statement of the 'self-help' ethos later popularised by Samuel Smiles. Nevertheless, together with virtually all the other senior engineers of his period, Brunel was persuaded to attend the Commission and to give oral evidence, and there is no reason to believe that he disassociated himself from the Report made by the Commission.

The members of the Commission spent two years examining the issues involved very thoroughly, promoting a number of tests and receiving a large volume of evidence. When published in July 1849 the Report comprised over 400 pages, but its conclusions were cautious. It recognised that:

'On the whole, the art of railway bridge-building cannot be said to be in that settled state which would enable an engineer to apply principle with confidence.'

And with regard to laying down firm theoretical principles for railway construction it concluded:

'we are of opinion that any legislative enactments with respect to the forms and proportions of iron structures employed therein would be highly inexpedient.'[23]

In this form there was little that even Brunel could find objectionable about the Report. But his intuition was basically correct, because the work of the Commission did mark a very significant development in the relationship between engineers and government. For one thing, it did lead to a number of limited reforms which were immediately incorporated into the *Requirements* which the Railway Inspectors issued in order to guide engineers regarding standards to be expected before certificates of approval would be issued. Again, the Commission established the important principle that, in enquiries involving engineering structures, it was necessary to conduct experiments and to collate information from 'objective' scientific sources. Most significantly, however, it confirmed the principle of governmental interest and involvement in matters of engineering safety on the railways. While the process of rigorous testing of wrought-iron structures by Eaton Hodgkinson, William Fairbairn, and others, produced dramatic improvements in the quality of large bridges, as demonstrated in Stephenson's Britannia tubular bridge over the Menai Straits, and Brunel's Royal Albert Bridge over the Tamar, the quiet persistence of the Railway Inspectors brought continuing improvements in signalling, braking, and other running practices, so that accidents attributable to engineering inadequacies became comparatively rare. Governmental control was thus imposed gently but nonetheless effectively on the engineers of the British railway network in the second half of the nineteenth century.

All such measures, however, proved unable to prevent the most dramatic of all British railway disasters. On the night of 28 December 1879, the Tay Bridge collapsed in a severe storm, carrying a complete train with seventy-four people on board to their doom. This wrought-iron lattice-girder bridge, supported in eighty-five spans by cast-iron columns, was almost two miles long. It had been completed in September 1877 and approved by the Railway Inspectorate. Queen Victoria had crossed it in the summer of 1879 and had taken the opportunity of bestowing a knighthood on the engineer who had designed it, Sir Thomas Bouch. Bouch was an experienced railway engineer who had served under Locke and Errington and who was proud of his reputation for economy and for not exceeding his estimates. In his lack of any sort of theoretical training he represented the traditional style of British engineer, and the disaster revealed in an acute form the serious nature of the inadequacies of this style.

The collapse of Bouch's Tay Bridge caused a national sensation, and a Court of Enquiry was hurriedly appointed under the terms of the Regulation of Railways Act of 1871 in order to investigate it.[24] The three members were William Henry Barlow, President of the Civils; Colonel Yolland, then Chief Inspector of Railways; and the Chairman, Henry Cadogan Rothery, who was the Commissioner of Wrecks. Of the three, Rothery had least engineering expertise, but he was a perceptive and somewhat truculent Chairman who knew his own mind and was determined to express it. The Court took evidence from 120 witnesses and commissioned an engineer, Henry Law, to make a minute examination of the remains of the bridge, as well as having special tests made at David Kirkaldy's testing establishment in Southwark. The basic reason for the disaster became clear enough: the bridge had blown over due to intrinsic structural weaknesses. But the reasons for these weaknesses and the responsibility for them were more difficult to determine. The engineering members of the Court, Barlow and Yolland, took the view that the main structural defect was inadequate cross-bracing, and that insufficient allowance had been made for wind pressure. The theoretical deficiency on such a fundamental aspect of bridge design is surprising, but it appears that Bouch had depended upon some calculations made for him by the Astronomer Royal, Professor Airy, which had drastically under-estimated the wind forces likely to be encountered by a structure of this type. With such reservations about the best theoretical information available, the engineers were disinclined to censure Bouch too heavily. However, Rothery, in a separate Report, went further than his colleagues and placed the responsibility for 'the design, the construction, and the maintenance' of the bridge firmly upon the shoulders of Bouch.[25] This judgment certainly destroyed Bouch's reputation and it probably hastened his death at fifty-eight in October 1880. It did not do anything for the good reputation of the engineering profession either. Nevertheless, the voluminous evidence submitted to the Court served once more, like that of 1849, as a valuable research exercise, so that subsequent bridge work gained substantially from the hard lesson of experience. When the Tay Bridge was rebuilt and the even mightier Forth Bridge was constructed they were made strong enough to resist any gales that the east coast weather could hurl against them, and they remain in operation to this day.

Railway works were the outstanding field in which engineering skills generated new operational systems in the nineteenth century requiring recognition, definition, and regulation by government, and it was a field in which a new government inspectorate combined with Royal Commissions and Courts of Enquiry and a maturing professional organisation on the part of the engineers to produce a safe and viable system. There were adjacent fields of engineering developments in which similar patterns of institutional evolution occurred. Amongst these were the provision of urban water supplies, the supply of town gas, the sea-worthiness of merchant ships, the use of iron and steel in ship construction, the strength and safety of steam boilers, and the emission of waste from chemical works.[26] In most cases, the professional monitoring of these matters was superseded by some form of governmental control in the second half of the century, with the engineering profession adapting itself to providing expert advice to government as requested.

A rather similar process of adaptation occurred in relation to labour organisations. Mid-century engineers had taken a determined stand against the rise of trade unions. The Fairbairn family promoted a vociferous, though largely anonymous, opposition to the Amalgamated Society of Engineers in the great engineering lock-out of 1851-52,[27] and James Nasmyth prided himself on successful strike-breaking before deciding to retire early rather than accept the need to work with trade unions.[28] By the end of the century, however, professional engineers had become largely reconciled to the existence of powerful trade unions as a necessary aspect of large-scale industrial organisation.[29] In this field as elsewhere the individualistic ethos praised by Samuel Smiles was gradually transmuted into a professional ethos in which the engineers fulfilled the role of expert advisers assisting the agents of government to exercise an increasingly interventionist function, but one which avoided any need for engineers themselves to become politically committed.

If engineers were muted in their political opinions, they were positively reserved in their religious beliefs. In most instances of which we have any record, these views were orthodox and conformist, provided that traditional English Nonconformity and Scottish Presbyterianism are included in this category. James Watt, Thomas Telford, and many other Scottish engineers, imbibed a strong quality of Presbyterianism, even though their religious practice appears to have been sporadic. James Nasmyth has nothing significant to say about religious beliefs or practice in his *Autobiography*.[30] William Fairbairn, on moving to Manchester, became a member of the Unitarian congregation at Cross Street Chapel, where the wife of the Minister, Mrs Gaskell, appreciated his comments on her novels.[31] At least one engineer abandoned his practice to become a clergyman: this was Henry Thomas Ellacombe (1790-1885) who had worked as an Assistant Engineer to Marc Brunel before being ordained. He subsequently became Vicar of Bitton near Bristol and Rector of Clyst St. George in Devon, and acquired a reputation as an accomplished antiquarian.[32] Several other engineers such as I. K. Brunel, C. B. Vignoles and John Fowler, had sons who became clergymen or otherwise entered the service of the church, and as these sons sometimes wrote biographies of their fathers, as in the cases of Brunel and Vignoles, they went to some pains to identify the religious orthodoxy of their parents. Brunel's son, for instance, wrote of his father taking comfort in adversity 'from those higher sources of consolation on which it was his habit to rely', and there are some ambiguous references to prayer in Brunel's letters.[33] But there is no evidence of regular church attendance in the papers of I. K. Brunel, and it is difficult to imagine him accepting patiently the pulpit oratory of his period. There is an intriguing annotation in his handwriting made in the margin of a copy of Buckle's *History of Civilisation* which appears to show him as being critical of the utilitarian view of religious ethics represented by Helvetius.[34] However pointless the exercise, moreover, it is curious to speculate that Brunel died a month before the publication of *Origin of Species*, and to wonder how the most intelligent and articulate of British engineers would have responded to the Darwinian bombshell had he lived long enough.

As so often elsewhere in this review of engineering ideas and beliefs, we lack sufficient hard facts to make a completely convincing reconstruction of religious attitudes. But the lack of evidence is itself indicative of a profoundly orthodox and conformist ethos, in so far as it suggests that there was nothing unusual requiring attention. While scientists of the period included amongst their number odd sectarians like Michael Faraday, fundamentalists such as Philip Gosse and Hugh Miller, and even spiritualists such as Sir William Crookes, it is difficult to identify any engineers with unconventional religious views. Locke offended Scots sabbatarianism by wanting to run passenger trains on Sundays, but that tells us nothing about Locke's own religious views.[35] Francis Fox insisted on a Sunday rest-day for all his workers and promoted social welfare in a non-denominational way, but was otherwise completely orthodox in his views.[36] Alfred Ewing, as the son of a minister of the Free Church of Scotland, never felt the need of some sons of the manse to cast off the restraints of religious conformity.[37] The tone of many presidential addresses in the *Transactions* of the engineering institutions was frequently rather sanctimonious in representing a conventional sort of religiosity, but behind this smokescreen the only persistent theme of a belief shared by all engineers in the nineteenth century was that of an assumption of beneficent progress. This, of course, was not distinctively an engineering phenomenon, as the belief was widely shared in the period, but in so far as engineering achievements seemed to have confirmed and justified progress, engineers could be regarded as having had a particular vested interest in it, and they suffered a specially acute ideological disillusionment with the weakening of belief in progress after 1914.

A belief in progress should certainly be regarded as an aspect of nineteenth century religion, but it was also an assumption which applied to the whole field of intellectual activity in the period. Engineers were only slightly less reticent about their beliefs in this area than they were in politics and religion, so that negative evidence is again sometimes more impressive than any positive indications of commitments, and for the most part engineers made only modest marginal contributions to British intellectual life. This can be illustrated by reviewing the contributions of engineers to science, literature and the arts.

With the close links between engineering and science which we now take for granted, it might have been expected that there would have been some distinctive engineering contributions to science in the nineteenth century. Many leading engineers from Smeaton onwards had been Fellows of the Royal Society, and Smeaton himself had established his reputation as a scientist before he became widely known as a civil engineer. His paper of 1759 - 'An Experimental Enquiry concerning the Natural Powers of Water and Wind to turn Mills...' was a pioneering study and set the pattern for much subsequent work of engineering science which developed model-testing techniques.[38] However, few leading engineers in the nineteenth century found time to prepare papers for the Royal Society. I. K. Brunel, for instance, although a Fellow from 1830, appears to have taken little part in any proceedings, and engineering matters generally received little attention from the Society in this period.[39]

Over against this paucity of scientific activity it should be remembered that several engineers distinguished themselves as amateur scientists, particularly in astronomy. Smeaton maintained careful astronomical observations from his Yorkshire home at Austhorpe.[40] Vignoles went to great lengths to try to observe solar eclipses.[41] Nasmyth made exquisite drawings of the lunar landscape which he studied through his telescope.[42] And

Charles Parsons inherited something of the enthusiasm of his father, Lord Rosse, for making and using large telescopes.[43]

While such interests give a useful clue to the intellectual activity of engineers, it is doubtful whether they helped greatly in advancing the frontiers of knowledge, but in the new science of thermodynamics the contribution of engineers was more fundamental. In a sense, they provided the experience and basic data out of which thermodynamics emerged, by stimulating speculation about the physics of the steam engine. Sadi Carnot set out to elucidate the relationship between heat and work in the high pressure steam engine, and the principles could then be extended to all sorts of heat engine and to the behaviour of the universe. Amongst the many international figures, at least one British engineer, W. J. M. Rankine, to whose creative work as Professor of Engineering at the University of Glasgow we have already given attention, took part in this exciting intellectual development.[44]

Other British engineers engaged actively in the discussions of the British Association for the Advancement of Science, established in 1831, usually on themes with a strong practical application.[45] John Scott Russell developed his 'wave line' theory of ship construction in papers to the Association, having carried out scale model tests with various shapes of hulls on canals, and I. K. Brunel argued with Professor Dionysius Lardner about the feasibility of transatlantic steam navigation and the safety of the Box Tunnel at meetings of the Association.[46] Brunel also argued with Professor William Buckland about the geological soundness of the Box Tunnel, and on both matters concerning the tunnel his practical wisdom proved to be more reliable than the more theoretical arguments of his 'scientific' opponents.[47] As far as Russell's research was concerned, subsequent work on ship model-testing by William Froude achieved a convincing reconciliation between scientific theory and practice.[48]

Charles Babbage was one of the pivotal figures of mid-nineteenth century intellectual life. Lucasian Professor of Mathematics at Cambridge and scourge of his fellow-members of the Royal Society for their narrow views of science, he helped to launch the British Association in order to popularise science and he expressed a keen interest in practical matters of engineering and manufacturing, particularly in relation to the construction of his proto-computer, the 'difference engine', which proved to be too complicated for the mechanical techniques of his time. Babbage's book, *On The Economy of Machinery and Manufactures*, published in 1832, did much to promote the systematic rational discussion of engineering. It also made an important contribution to the confident progressive ethos of the period.[49]

Although not insignificant, the contributions of nineteenth century engineers to the practical sciences were thus low-key and generally marginal, and as far as the more theoretical sciences and the life sciences of biology and zoology were concerned, they were virtually non-existent. The great evolutionary debate which stemmed from the publication of Charles Darwin's *Origin of Species* in 1859 left them comparatively unscathed, although there was one significant exception in the publication of a review of *Origin* by the engineer Fleeming Jenkin, who developed a serious critique of Darwin's theory on the issue of blending inheritance, arguing that any random genetic 'improvement' would be blended out over a few generations.[50] The point was sufficiently important to cause

Darwin considerable perplexity,[51] but otherwise the engineers made little contribution to the debate which characterised so much of Victorian intellectual life.

In the social sciences, also, one engineer proved the exception to demonstrate the general rule that engineers had little contribution to make. This was Herbert Spencer (1820-1903), who began his career as a railway engineer but developed a theory of social evolution according to the survival of the fittest even before Charles Darwin's views had become well known, and went on to become one of the most eminent and controversial social commentators of his time. His great work, *The Synthetic Philosophy*, began in 1860 and was completed in 1896, running to many volumes. Spencer did as much as Marx to popularise the idea of 'social engineering' in the nineteenth century.[52]

In other fields of intellectual endeavour, such as literature, art, architecture and music, the direct contribution of engineers was equally slight, although some of them did perform a useful role as patrons of the arts. Many engineers were capable of fluent eloquence in professional reports, in the presentation of engineering cases to Parliamentary and other committees, and in private correspondence, but there are few recorded instances of any other form of literary distinction. Thomas Telford preferred to keep his poetry more or less to his personal friends, and he had no professional imitators as an engineer-poet.[53] His attempt at autobiography was a literary failure,[54] and few other engineers managed to write a presentable autobiography, apart from Nasmyth, whose work was virtually rewritten by Smiles.[55] R. E. Crompton wrote some interesting personal reminiscences,[56] and Sir Alfred Ewing published some lively essays,[57] but on the whole engineers seemed reluctant to express themselves in print, and when they did so it was usually with no great distinction. There was some good engineering journalism, especially associated with Zerah Colburn (1832-70), an American who came to Britain to edit both *The Engineer*, founded in 1856, and *Engineering*, founded in 1866. These journals established a secure reputation for sound and serious commentary on engineering topics, but they never aspired to literary pretensions.[58]

Aristocrats and gentry, clergymen and civil servants, politicians and journalists, housewives and architects, were all represented amongst the successful novelists of the nineteenth century, but not - so far as we have been able to discover - engineers. It is true that R. L. Stevenson was trained as an engineer, but he rejected the career as a young man as soon as he could throw off parental control, so he cannot really be regarded as an exception.[59] Nor did engineering receive much attention in the novels of the period, despite the tremendous vogue for this art form. The best example amongst novels of any substance is probably Anthony Trollope's story, *The Claverings*.[60] In this the hero, Harry Clavering, is described as the fellow of a Cambridge college who decides to reject a career in the church and to re-train as a civil engineer. He tells his family:

'You see I could have no scope in the church for that sort of ambition which would satisfy me. Look at such men as Locke, and Stephenson, and Brassey. They are the men who seem to me to do most in the world. They were all self-educated, but surely a man can't have a worse chance because he has learned something.'[61]

It has to be said, however, that engineering events such as the construction of the London Underground and the projection of a railway in Russia figure only fleetingly and peripherally in the plot, which is involved primarily with complicated courtship rituals of a conventional Victorian form. In the denouement, moreover, our hero stands to inherit a baronetcy and the engineering career is abandoned in favour of a life of leisure. The engineer, later to become his brother-in-law, at whom Clavering initially looks critically because 'he had an odious habit of dusting his shoes with his pocket-handkerchief', says of this arrangement:

'Providence has done very well for him... but Providence was making a great mistake when she expected him to earn his bread.'[62]

Far from magnifying the profession of the engineer, the novel thus gives support to the thesis that Victorian gentlemen sought satisfaction in rural pastimes rather than in creative labour.

Engineers did manage to provide some patronage for the arts. In redecorating his Duke Street house I. K. Brunel decided to commission several leading artists of the day to provide him with paintings. His choice of artists, however, was surprisingly conventional - Landseer, for example, was invited, but none of the 'revolutionary' Pre-Raphaelite Brotherhood - and the theme selected was that of scenes from Shakespeare, so that even Landseer ended up painting a subject which was very unusual for him - a murky scene depicting Titania and Bottom from *A Midsummer Night's Dream*.[63] Some decades later, Landseer was entertained at the Highland home of Sir John Fowler, on his Braemore estate, as were J. E. Millais and other artists.[64] James Nasmyth came from a family of artists and could do very competent drawings himself, and several engineers had their portraits painted by eminent artists.[65] Engineers also gave commissions to architects for houses and gardens, the outstanding example being the house at Cragside in Northumberland which Norman Shaw designed for Sir William Armstrong. No expense was spared for the materials which went into this extraordinary mansion, now owned by the National Trust.[66] The house planned by Brunel for his Watcombe estate near Torquay was never built, but the garden on which he took expert advice and supervised much tree planting has now grown to maturity, although much overgrown and neglected.[67]

Engineers gave some patronage to music. I. K. Brunel courted and married Mary Horsley, whose family cultivated the company of Mendelssohn and other musicians, and the Brunels maintained these associations.[68] There is no record, however, of any distinctive instrumentalists, vocalists, or composers amongst Victorian engineers, although John Scott Russell and his family had a lively 'music room' at their Sydenham home where Arthur Sullivan was a regular visitor.[69] It should also be noted that Sir George Grove (1820-1900), author of the magisterial four-volume *Dictionary of Music and Musicians*, had been trained as a civil engineer and had supervised the erection of lighthouses in the West Indies in the 1840s.[70]

Engineers generally were not tempted to take a lead in the artistic exploration of the new industrial society which they were doing so much to create. While Turner experimented with representations of steam locomotives and ships in his pictures, and Tennyson and Kipling made cautious use of engineering imagery in their poetry, the engineers made little artistic response to industrialisation.[71] They were puzzled but un-

moved by aesthetic criticisms of their structures, which grew in volume with the multiplication of wrought-iron girder bridges in the second half of the nineteenth century. It
was left to a later generation of art critics to idealise their work as the 'functional tradition'.[72] As far as the engineers responsible for the new railway bridges across the Thames
in the heart of London were concerned, they were simply delivering the best and most
reliable structures within the budget available to them.[73] Even Brunel, who has rightly
not been regarded as a cheese-paring engineer, incurred the silent disapproval of Queen
Victoria for the severe functionalism of the girder bridge which he erected over the Dee
at the entrance to her Balmoral estate.[74] And it must be said that the engineers have been
proved right in many cases where their aesthetic judgment has been called in question.
For instance, the attack by William Morris on the Forth Bridge when it approached completion in 1889 now seems misplaced:

> 'There never would be an architecture in iron, every improvement in machinery being uglier and uglier,
> until they reach the supremest specimen of all ugliness - the Forth Bridge.'[75]

Considering that the graceful steel cantilevers of this splendid structure are now properly regarded as displaying the best features of nineteenth century engineering, the aesthetic criticism of Morris has come to appear bizarre.

In their lack of intellectual distinction outside their own specialist field, the engineers
of our period can be seen as non-demonstrative ideological conformists, anxious to identify themselves with the prevailing mores and conventions of society. This is strikingly apparent in the unanimity with which leading engineers adopted the habits and life-styles
of successful professional men, especially after the middle of the century. Up until then
there had been some flexibility of attitudes, so that Telford had lived happily for most of
his professional career in the Salopian Coffee House in Charing Cross, not acquiring a
home of his own until 1821, when he was already sixty-four.[76] With a similar indifference
to convention, Telford, George Stephenson and Robert Stephenson all refused the honour of knighthoods, although Telford had a Swedish knighthood thrust upon him and
George Stephenson accepted a Belgian knighthood. But with Smiles' adoption of the engineering profession for his paragons of self-help and the Victorian virtues, the attitudes
of the profession became more conventionally stereotyped. Honorific titles became welcome indications of social acceptance, and could be regarded in part as rewards for social conformism. Thus the engineering profession, which had begun in the middle of the
eighteenth century as a very motley group of men, mostly from extremely humble backgrounds, had within a century won recognition of its members as being not only
gentlemen, as would be required of any acceptable professional body, but as aspirants to
social gentility, with all the accoutrements of titles, estates, and the way of life to go with
them. This represented a very significant hardening of engineering attitudes, emphasising social conformity and obliterating any expression of radical or heterodox opinions.

The earliest engineering knighthoods in Britain had been awarded in the first half of
the seventeenth century, when Sir Hugh Myddelton and Sir Cornelius Vermuyden were
both so honoured.[77] After that, however, there was a long gap before further titular honours were bestowed on British engineers. None of the eighteenth century pioneers of the
new profession achieved the accolade. John Rennie is said to have refused it, but his son

accepted it readily enough on the completion of his father's design for the new London Bridge in 1832. Sir Marc Brunel was knighted in 1841 on the completion of the Thames Tunnel. Then from the middle of the century there developed a steady flow of engineering titles. The Great Exhibition of 1851 produced three: Sir William Cubitt, as chairman of the management committee; Sir Joseph Paxton, the gardener from Chatsworth who became the designer of the Crystal Palace; and Sir Charles Fox, of the contractors Fox and Hendersons, who built it. Then Sir William Armstrong, the hydraulics and armaments engineer of Newcastle-upon-Tyne, received a title in gratitude for his improvements in gun-making, even though the conservatism of the naval establishment caused the nation to be reluctant to make use of them. Sir Daniel Gooch was made a baronet in 1866 - not for his distinguished work in making broad-gauge locomotives for the GWR but for his success in cable-laying. Sir Thomas Bouch was knighted on completion of his ill-fated Tay Bridge in 1877, and undeterred by the tragic end of that story the nation acted promptly on the opening of the Forth Bridge in 1890 by rewarding both Sir John Fowler with a baronetcy and his partner Sir Benjamin Baker with a knighthood. Fowler had already received a knighthood in 1885 for services to the government in connection with his engineering experience in Egypt.[78] Sir John Coode had meanwhile been knighted in 1872 for his harbour works at Portland and elsewhere; Sir Joseph Bazalgette had been honoured for his completion of London's main drainage and sewage-disposal system in 1874, and Sir Joseph Whitworth's services to the machine tool and armaments industries had been recognised in 1869, while his great rival Armstrong had been elevated to the peerage - the first engineer to receive this distinction. Two other machine tool-makers, the Fairbairn brothers, both received the accolade, although the younger brother, Sir Peter Fairbairn, received his title largely for political services in Leeds in 1858, while the elder brother Sir William was awarded a baronetcy in 1869 at the age of eighty.[79]

It is notable that railway engineers as a whole did not do well in these honours stakes: it was left to the twentieth century to honour Sir John Aspinall, Sir Nigel Gresley, and Sir William Stanier. Aspinall, who was knighted in 1917, has been described as one of the class of 'engineering managers',[80] and it is significant that several of those engineers already named such as Gooch and Fowler fall firmly into this category, making it apparent that what was being rewarded in all these cases was not mere engineering talent but also a capacity for management and social accomplishment. The same tendency is apparent in the history of the five Tangye brothers who moved from Cornwall to Birmingham in the 1850s and established there one of the greatest engineering firms in the second half of the nineteenth century. But of the five, it was the fourth and the one with the least mechanical skill who became the manager and was awarded the knighthood as Sir Richard Tangye in 1894 for 'many public services and steadfast political faith'.[81]

Whether titled or not, many nineteenth century engineers acquired substantial wealth from their professional practice, and this assisted materially in endorsing their prestige as gentlemen who had achieved social respectability. Armstrong at Cragside and Fowler at Braemore were outstanding examples of engineers who consolidated their professional success with large estates and splendid mansions, but there were many others. Brunel did not live to enjoy his Watcombe estate, but he chose the site with great care and devoted much attention to planning it even when he was chronically overworked with his business commitments, becoming involved in problems with the siting of the local gasworks and the conduct of the parish clergyman.[82] Archibald Sturrock, Locomotive Superintendent of the Great Northern Railway, retired in 1866 at the age of fifty in order to live the life of a country gentleman until 1909, described by one historian as 'a sort of synthetic John Bull'.[83] He was by no means unusual, and there were other sorts of gentlemanly pursuits which could be adopted. Robert Stephenson, for example, spent lavishly on his luxury yacht, the *Titania*, and used it for touring to Scandinavia and the Mediterranean as well as for general relaxation.[84]

Successful gentrification was inimical to the dynastic continuity of the engineering profession. There had been some remarkably versatile dynasties during the nineteenth century, such as the Jessops, the Rennies, and the Mylnes. Robert Stephenson had followed his father into engineering, and although he had no children of his own the dynasty was carried on through his cousin George Robert Stephenson who, however, showed less aptitude or interest in engineering than his predecessors. With the Brunels, the engineering talent spanned three generations, from Marc Isambard through Isambard Kingdom to Henry Marc Brunel, but the family then became respectably gentrified and lost to engineering. The Stevensons, a gifted family of Scottish lighthouse builders, established their reputation with Robert (1772-1850) and continued in the second generation with Alan (1807-65), David (1815-86), and Thomas (1818-87) all practising as engineers. The most famous member of the third generation was Thomas's son Robert Louis (1850-94), whose father assumed that any son of his would naturally become an engineer and trained him for this career, only to be disappointed when his son took to writing novels.[85] In none of these cases was gentrification a paramount feature in breaking the dynastic continuity of the profession, but it contributed to this process, and the effect was much more pronounced in other families. When Sir John Fowler died in 1898, for example, one of his four sons had become a clergyman and another a farmer, and while a third was employed in some Spanish mines it does not appear that any of them had been trained for a career in engineering.[86] Similarly with the Rendels, the Cubitts, and other successful nineteenth century engineering families: the younger generations tended to disperse rapidly into other professions and occupations, giving some support to the 'haemorrhage of talent' thesis applied by some scholars to the performance of the British economy at the end of the nineteenth century.[87]

The achievement of titles, wealth and estates all provide indices of the status of the engineers as a professional group which consisted of gentlemen and which was being assimilated into the gentry. Another such index is the monuments by which the lives of the engineers were commemorated. Two nineteenth century engineers were accorded the honour of burial in Westminster Abbey, that mausoleum of national genius: Thomas Telford and Robert Stephenson lie within a few feet of each other and are commemorated by plaques on the floor of the nave. I. K. Brunel and Joseph Locke are both buried in Kensall Green Cemetery, but the friends and family of Brunel financed a memorial window in the south aisle of the nave of Westminster Abbey, designed by Norman Shaw, the fashionable architect who built Cragside for Armstrong. Opposite this, the north aisle of the Abbey forms a sort of 'Engineers' Corner' with a row of memorial windows commemorating Richard Trevithick (the only one with any visual representation of engineering structures, depicting four key inventions), Frederick Henry Royce, Charles Algernon Parsons, John Wolfe Barry, Benjamin Baker, and Baron Kelvin of Largs. The choice seems fairly random and may have depended primarily on the devotion and resources of family and friends. Brunel was also commemorated with a statue by Carlo Marochetti on the Embankment in London, and by the completion of the Clifton Suspension Bridge in Bristol.[89] There are local memorials to several other engineers, including a statue of Trevithick in Camborne, of George Stephenson in Newcastle and of Joseph Locke in Barnsley. Although the nation is not over-endowed with monuments to its great engineers, there are thus sufficient to demonstrate that they have made a significant impression and have been held in high esteem by a grateful public.

One other point should be considered in assessing the close integration of the nineteenth century engineering profession into a network of social orthodoxy and conformity - the role of the professional institutions. We have pursued at length the role of these institutions in catering for the increasingly diverse and specialised professional needs of engineers, and their role as means of organised self-education, but it is apparent that they also performed for their members a valuable function as social clubs. The value of this service derived from the fact that gentlemen's clubs were an important part of the social milieu into which the engineers aspired to move, so that the possession of a club affiliation was a necessary step in establishing social status. As Anthony Sampson pointed out some years ago:

'The club is a pervading image among British institutions... The point of a club is not who gets in, but who it keeps out. The club is based on two ancient British ideas - the segregation of classes, and the segregation of sexes.'[90]

The development of engineering institutions coincided closely with the remarkable flowering of gentlemen's clubs in London, and it was thus natural to conceive the institutions as clubs for engineers, providing the facilities for meeting and dining which gentlemen expected of their clubs. By conforming to the pattern of gentlemen's clubs, the professional engineers acquired in their institutions one of the vital outward and visible forms of gentlemanly behaviour.[91]

Economic historians have long been familiar with the problem of the comparative decline of Britain in the increasingly competitive world trading situation at the end of the nineteenth century, and the engineers have not escaped criticism for some responsibility in this situation, with their tenacious defence of apprenticeship and their tendency towards gentrification.[92] This argument has recently been reinforced from the point of view of social and literary history by an American scholar, Martin Wiener, who has argued that, from the middle of the nineteenth century, the influential professional classes in Britain tended to turn their minds away from industrialisation and its problems towards a rural idyll of a cottage in the country and a society dominated by traditional values and relationships. Wiener presents a formidable mass of evidence to demonstrate that British social, political and literary opinion has been strongly influenced by such aspirations, and the implication that this has diverted talent from practical problems of making industrial society work better and more effectively is not easily refuted.[93] To some extent, the evidence of engineering ideas and beliefs reviewed in this chapter, fragmentary as it has necessarily been, tends to confirm the Wiener analysis. Nevertheless, the case is over-stated, and is less convincing when it is examined in detail. For instance, I. K. Brunel is cited by Wiener as an example of an engineer who decided to seek a higher social status for his sons by sending them to public school, while the fact that one of these sons served an engineering apprenticeship and became a successful engineer in his own right is ig-nored.[94] Evidence from the anonymous mass of engineers is, by definition, hard to come by, but the inference to be drawn from studies of engineering firms and railway work-shops suggests that most engineers were content to get on with their own jobs to the best of their ability without dwelling on possibilities of upwards social mobility and gentrifi-cation.[95] It is necessary, therefore, to guard against the consequences of historical 'whig-gery' in this matter, whereby a case is demonstrated by selecting only favourable evidence and ignoring the rest. Wiener is not completely innocent of this procedure.

Our biggest problem in considering the ideas and beliefs of the engineers has been the lack of specific evidence, but it is not unreasonable to infer that the consequent an-onymity of the engineers derives in part from a high degree of job satisfaction. There are many indications that the British professional engineers worked very hard in the nine-teenth century: several burnt themselves out with overwork and even of those who stood the strain there is plenty of evidence that they would work for exceedingly long hours at particular periods, as when trying to meet deadlines for the Parliamentary calendar.[96] This work ethic may have been partly religious and may even have been partly related to emotional insecurity, but it seems fair to imply that most engineers liked their work and were prepared to devote most of their lives to it. The fact that this left them little time or inclination to develop political or social interests, to indulge flights of philosophical or theological speculation, or even to attend to their families or to take reasonable holidays, goes far towards explaining the curious reticence of the engineers in any matters not of immediate professional concern to them. Unlike lawyers or doctors, or even craftsmen such as masons and cobblers, engineers have been persistently committed to practical deeds rather than words. Their images have been physical rather than literate, and their visions have been realised in iron and glass and stone rather than in printed words and works of art executed in the study or the studio.[97] Engineers have been men with con-crete rather than speculative imaginations, so that for their substantial monuments it is necessary to look about us rather than consult any written record. Their considerable in-tellectual faculties have generally been subordinated to the overriding needs of the con-structional task of the moment. For the rest, unless deeply disturbed by the interference of outside influences, they have been content to go along with majority opinion and to reflect the consensus. Their work has frequently depended on the state and large corpor-ations, so that their own interests have dictated conformity and orthodoxy as the most appropriate intellectual postures. In this, interest has coincided with temperament to frame engineers in the mould of organisation men.

Notes

1. Sir Daniel Gooch, *Memoirs and Diary*, edited by R. B. Wilson, David and Charles, Newton Abbot, 1972, p. 345.

2. William Wordsworth, *The Prelude*, Book II, line 108.

3. L. T. C. Rolt, *Isambard Kingdom Brunel*, 1957, p. 63, comments on Brunel's political sympathies. Lady Celia Noble, *The Brunels - Father and Son*, 1938, 184-5, wrote of Brunel having 'dashed over to Paris for the pleasure of seeing the famous barricades on the Boulevards' in April 1848, but at home he enlisted as a special constable in Westminster in the wake of public alarm about possible Chartist riots.

4. Memoranda by I. K. Brunel on both government control of engineering works (13 March 1848) and the Patent Laws (1 July 1851) are quoted in I. Brunel, *The Life of Isambard Kingdom Brunel*, 1870, reprinted 1971, 486-498.

5. L. T. C. Rolt, *George and Robert Stephenson*, 1960, 324-427.

6. Charles Walker, *Joseph Locke*, 1975, 40-42. This is a slight though useful booklet: for a fuller account, see N. W. Webster, *Joseph Locke - Railway Revolutionary*, 1970.

7. T. Mackay, *The Life of Sir John Fowler*, 1900, 120, 121, and 345.

8. *Mechanical Engineering News*, 138, June 1984.

9. Charles Hadfield and A. W. Skempton, *William Jessop, Engineer*, 1979, p. 253.

10. Francis Fox, *Sixty-three years of Engineering*, 1924, p. 273.

11. *John Brunton's Book*, introduced by J. H. Clapham, 1939.

12. See *DNB* entry for Charles George Gordon.

13. K. H. Vignoles, *Charles Blacker Vignoles - Romantic Engineer*, Cambridge, 1982. See also *DNB* entry for Vignoles.

14. See O. J. Vignoles, *Life of Charles Blacker Vignoles*, 1889, p. 328.

15. B. H. Becker, *Scientific London*, 1874, reprinted 1968, p. 112.

16. Mackay, *op. cit.*, p. 183.

17. See, for instance, the reference to the Rifle Volunteers in Brian Reed, *Crewe Locomotive Works and its Men*, Newton Abbot, 1982, p. 211.

18. Mackay, *op. cit.*, p. 314: 'Mr. William Morris, a great handicraftsman and man of genius, who unfortunately allowed his fine artistic sense to be obscured by a disordered political imagination'.

19. Henry Parris, *Government and the Railwaymen in Nineteenth Century Britain*, 1965.

20. See above, Chapter Four.

21. The six commissioners appointed were George Rennie and William Cubitt, both engineers; Captain Henry James, a Royal Engineer; Professor Robert Willis, an academic; Eaton Hodgkinson, a mathematician; and the Chairman, Lord Wrottesley. The Secretary was Lt. Douglas Galton, RE.

22. I. Brunel, *op. cit.*, p. 487: see note 4 above.

23. *The Report of the Commission on the Application of Iron to Railway Structures*, P.P., 1849, (1123), xxix xvi and xviii.

24. For the Tay Bridge, see above, Chapter Six.

25. *Report of the Court of Inquiry ... upon the circumstances attending the fall of a portion of the Tay Bridge on the 28th December 1879*, P.P., 1880, (2616), xxxix, the report of Mr Rothery, p. 44 para. 120.

26. See, for instance, R. M. McLeod, 'The Alkali Acts Administration, 1863-84: The Emergence of the Civil Scientist', in *Victorian Studies*, 9, 1965.

27. W. Pole (ed.), *The Life of Sir William Fairbairn*, 1877, reprinted 1970, 322-327. Fairbairn's son, later Sir Thomas Fairbairn, contributed an acerbic series of letters to *The Times* under the pseudonym 'Amicus' which were very critical of the unionists.

28. *Royal Commission on Trades Unions, Tenth Report*, P.P., 1868, xxxix, questions 19,222 and 19,223: cited in J. A. Cantrell, *James Nasmyth and the Bridgewater Foundry*, Manchester, 1984.

29. J. B. Jefferys, *The Story of the Engineers 1800-1945*, 1945.

30. S. Smiles (ed.), *James Nasmyth, Engineer - An Autobiography*, 1885.

31. Pole (ed.), *op. cit.*, 456-462.

32. For Ellacombe, see *DNB* entry. Also R. Beamish, *Memoir of the Life of Sir Marc Isambard Brunel*, 1862, 115-124.

33. I. Brunel, *op. cit.*, p. 516.

34. I am indebted to Dr Mark Gray for drawing my attention to the annotations in his copy of Buckle's work. This copy bears the book-plate of I. Brunel and a pencil note refers to the annotations by his father.

35. C. Walker, *op. cit.*, p. 41.

36. F. Fox, *op. cit.*, 242-248.

37. Sir Alfred Ewing, *An Engineer's Outlook*, 1933 - a pleasant collection of essays and papers, including one, IX, 'A Lay Sermon', 240-247. See also A. W. Ewing, *The Man of Room 40 - The Life of Sir Alfred Ewing*, 1939, p. 244 - 'His religion ... was not only conventional, but practical'.

38. Norman Smith, 'Scientific Work', chapter II in A. W. Skempton (ed.), *John Smeaton, FRS*, 1981.

39. I. Brunel, *op. cit.*, p. 516 note 1. The Royal Society was criticised in this period, amongst other things, for its lack of practical application - see C. Babbage, *Reflections on the Decline of Science in England and on some of its causes*, 1830, reprinted 1971.

40. S. Smiles, *Lives of the Engineers*, 1862, reprinted 1968, 2, 75-76.

41. K. H. Vignoles, *op. cit.*, 139-140.

42. S. Smiles (ed.), *Nasmyth*, reproduces several of these drawings as plates.

43. Rollo Appleyard, *Charles Parsons - His Life and Work*, 1933, Chapter X - 'The Astronomer's Son'.

44. D. S. L. Cardwell, *From Watt to Clausius - The rise of thermodynamics in the early industrial age*, 1971, especially Chapter 8 - 'The New Science'.

45. J. Morrell and A. Thackray, *Gentlemen of Science - Early Years of the British Association for the Advancement of Science*, 1981.

46. Ibid., 473-474.

47. R. A. Buchanan, 'Science and Engineering - A Case Study in British Experience in the mid-nineteenth century', in *Notes and Records of the Royal Society*, 32, 2, March 1978, 215-223.

48. See above, Chapter Nine.

49. This work of Babbage is highly regarded as a contribution towards the history of technology in Jan Sebestik, 'The Rise of the Technological Science' in *History and Technology*, 1, 1, 1983, 25-43.

50. Fleeming Jenkin, review of *Origin of Species*, in *North British Review*, 46, 1867, 277-318. My colleague Professor David Collard has drawn my attention to the fact that Jenkin has also been taken seriously by economic theorists - see J.A. Schumpeter, *History of Economic Analysis*, 1955, 1986 edition 837-40, and Mark Blaug, *Economic Theory in Retrospect*, Cambridge, 1962, 3rd edition 1978, especially p. 315 and p. 341.

51. This point is made by D. R. Oldroyd, *Darwinian Impacts*, 1980, 135-8.

52. Ibid., 204-211 for a brief account of Spencer's work. As a young man he had worked as an engineer on the London to Birmingham and Birmingham to Gloucester railways.

53. L. T. C. Rolt, *Thomas Telford*, 1958, Pelican ed. 1979, 29-30, 44, 74-5, and 203.

54. J. Rickman (ed.), *Life of Thomas Telford, Civil Engineer, written by himself...*, 1838. Rolt, *Telford, op. cit.*, says of this: 'as a personal document it is valueless' - p. 207.

55. S. Smiles (ed.), *Nasmyth, op. cit.*

56. R. E. Crompton, *Reminiscences*, 1928.

57. A. Ewing, *op. cit.*, note 37 above.

58. Both journals have continued publication to the present.

59. R. L. Stevenson's contribution was to engineering biography, in his *Memoir to Fleeming Jenkin*, 1887, and *Records of a Family of Engineers*, 1894.

60. A. Trollope, *The Claverings*, 1867.

61. World Classics edition, Oxford, 1986, p. 21.

62. Ibid., p. 514.

63. Richard Ormond, *Sir Edwin Landseer*, Exhibition Catalogue, 1981, 189-190 - the other artists involved were Sir Augustus Wall Callcott, Charles West Cope, Thomas Creswick, Augustus Leopold Egg, Charles Robert Leslie, Daniel Maclise, and Clarkson Stanfield.

64. Mackay, *op. cit.*, p. 328.

65. Sir John Fowler's portrait, for instance, was painted by Sir John Millais for the Civils - it is reproduced in Mackay, *op. cit.*, opposite p. 184.

66. Cragside was the first house in the world to be lit by hydroelectric power, on a system developed by Armstrong.

67. Rolt, *Brunel, op. cit.*, p. 235. The gardens have recently been brought to public attention by the energetic advocacy of Mr. Geoffrey Tudor - see *Brunel Society Newsletter*, Brunel Technical College, Bristol, no. 39, February 1986.

68. Rolt, *Brunel*, 91-92.

69. George S. Emmerson, *John Scott Russell*, 1977, Chapter 12 - 'The Sydenham Set', 259-269.

70. Ibid., p. 260. See also *DNB* entry for Sir George Grove.

71. Herbert L. Sussman, *Victorians and the Machine - The Literary Response to Technology*, Cambridge, Mass., 1968.

72. J. M. Richards, *The Functional Tradition in early industrial buildings*, 1958.

73. *Engineer*, 21, 4 May 1866, 315: see also 7 April 1865 and 27 July 1866; *Engineering*, 1, 11 May 1866, 309.

74. R. A. Buchanan and S. Jones, 'The Balmoral Bridge of I. K. Brunel' in *Industrial Archaeology Review*, 4, 3, Autumn 1980, 214-226.

75. Mackay, *op. cit.*, quoted p. 314.

76. Rolt, *Telford*, p. 198.

77. See above, Chapter 2.

78. Mackay, *op. cit.*, p. 274.

79. Pole (ed.), *Fairbairn*, p. 396 and p. 450.

80. C. Hamilton Ellis, *Twenty Locomotive Men*, p. 193.

81. S. J. Reid, *Sir Richard Tangye*, 1908, p. 249.

82. Celia Noble, *The Brunels*, 194-6. For correspondence regarding the gas works, see *Private Letter Books* for 5 June 1854, 25 February 1856, 22 October 1857, and 1 December 1857.

83. Ellis, *op. cit.*, p. 73.

84. Rolt, *Stephensons*, 328-333. He was frequently joined by his friend and fellow-engineer G.P. Bidder - see E. F. Clark, *George Parker Bidder - The Calculating Boy*, Bedford, 1983, 93-95.

85. Craig Mair, *A Star for Seamen - The Stevenson Family of Engineers*, 1978, p. 198.

86. Mackay, *op. cit.*, is curiously evasive about the details of Fowler's family.

87. For a discussion of the Wiener thesis, see below, and note 93.

88. The window, originally in the north aisle according to I. Brunel, *op. cit.*, p. 520 note 1, is partially obscured by monumental masonry.

89. I. Brunel, *op. cit.*, p. 520.

90. Anthony Sampson, *Anatomy of Britain*, 1962, 66-67.

91. R. A. Buchanan, 'Gentlemen Engineers: the making of a profession', in *Victorian Studies*, 26, 4, Summer 1983, 409-429.

92. See above, Chapter Nine.

93. Martin J. Wiener, *English Culture and the Decline of the Industrial Spirit, 1850-1980*, Cambridge, 1981.

94. Ibid., p. 19.

95. This view is confirmed by a variety of local studies such as Brian Reed, *Crewe*, note 17 above.

96. F. R. Conder, *The Men who built railways*, edited by Jack Simmons, 1983 - first published in 1868 as *Personal Recollections of English Engineers*.

97. Brooke Hindle, *Emulation and Invention*, New York, 1981, especially Chapter 6 - 'The Contriving Mind', 128-142.

Chapter Eleven

Conclusion: Engineering and Society

The outbreak of the First World War is a convenient point at which to conclude a review of the origins and evolution of the British engineering profession in the eighteenth and nineteenth centuries. There are good reasons for not taking the account any further, of which the most substantial is the difficulty of assimilating an enormous quantity of material without upsetting the balance of the historical narrative and blurring the issues on which this book has attempted to focus. There are also important shifts of emphasis associated with the war of 1914-1918, regarding both the composition of British engineering and the relationship between engineering in Britain and developments in the rest of the world. In several significant respects, such as the emergence of new industries in electronics and petrochemicals and the growing importance of international factors, developments since 1918 have involved innovations which require a somewhat different treatment from that which has been appropriate to the earlier history of engineering. For our purposes, it will be adequate to dispose of these innovations in summary form, while seeking to identify the continuities of engineering history in the twentieth century. Such a perspective accords with a major conclusion of this study, which is that the pattern of professional engineering in Britain was firmly established in the years before 1914, and that however great the subsequent expansion of the profession it is to those years that we can most usefully turn to understand the current attitudes and problems of the engineers.

The emergence of professional engineering has mirrored the progress of industrialisation and the consequential social changes in Britain over the years since 1750. It was intimately involved in the major constructional works which distinguished the first phase of this process, and went on to become a prime agent in the development of the skills and industries which made Britain the workshop of the world in the nineteenth century. Engineers made an essential contribution to the establishment of a modern system of transport and communications, and to bringing civilised standards of urban services to the cities of Victorian Britain. In the process, they acquired a self-consciousness and collective identity which converted them from a group of diverse though talented individuals into a professional elite, as which they achieved confident expression in the Institution of Civil Engineers. From the granting of its Royal Charter in 1828, the Institution represented the engineering profession in Britain, giving it form and definition and providing a guarantee of social status and respectability. It offered only rudimentary controls over entry into the profession, but its emphasis on the function of mutual education compensated for the lack of formal higher education for engineers and the consequent continued reliance upon instruction through a form of apprenticeship which, despite its shortcomings, maintained a viable and flexible organisation.The profession subsequently showed rather less flexibility in assimilating new skills in the period after 1860 when developments in thermodynamics and the sciences of electricity and chemistry began to impinge on practical engineering, so that the reluctance within the profession to modify its well-tried and successful methods of training can be seen as one factor in the structural changes in British industry at the end of the nineteenth century and the beginning of the twentieth century which tempered entrepreneurial vitality.[1] Adjustments within the profession, and particularly the insistence on greater theoretical competence from its members as exemplified by appropriate courses of instruction and examination, had gone far towards removing this weakness by 1914. However, by that time the profession had accepted the proliferation of specialised institutions amongst its members to such an extent that the Institution of Civil Engineers had lost its ability to represent the whole profession without any alternative single voice having emerged. While growing steadily in numbers and economic importance, therefore, the engineering profession lost cohesion and unity, so that engineers came to feel that their social status had deteriorated. This problem has continued since the First World War, and there is still no clear indication of a solution to it.

In one obvious sense, the year 1914 was not a date of particular significance in the history of the British engineering profession: it represents only a momentary blip in the upward curve of membership of the profession. As we have established in this study, in 1914 there were about 40,000 accredited members of seventeen professional engineering institutions in Britain. By 1979, the total had risen to over 375,000, represented by the membership of the sixteen Chartered Institutions which were then corporation members of the Council of Engineering Institutions (CEI) together with those of nine with affiliate status.[2] The latter category included at least one - the North East Coast Institution of Engineers and Shipbuilders with 1,358 members - which was excluded from the national figures in 1914 as it was then classified as a regional association. There are several other changes in the composition of the lists of institutions, apart from the growth from seventeen to twenty-five in the number included in the review. For instance, the Institution of Automobile Engineers and the Institution of Locomotive Engineers have both been integrated into the Institution of Mechanical Engineers, so that they do not appear in the 1979 list. On the other hand, there have been several new foundations since 1918 of bodies which had more members in 1979 than the largest institution had in 1914, and the emergence of these new organisations is significant because it represents the growth-edge of the profession.

The Institution of Chemical Engineers, for example, was founded in 1922 and had a membership of 12,035 in 1979. While its first President, Sir Arthur Duckham, expressed some reservations about the desirability of establishing yet another technical society, he nevertheless accepted the need to 'give a definite status to the body of men who must be responsible for industrial development in the future, that is to say, the chemical engineers',[3] and the growth of the petrochemical industry in the years between the World Wars ensured a rapid expansion for the new Institution, which received a Royal Charter in 1957. Similarly, the Institution of Electronic and Radio Engineers underwent dynamic growth from its foundation in 1925 to incorporation by Royal Charter in 1961 and a membership of 13,029 in 1979.[4] It was greatly stimulated by the use of radio in the First World War and by the introduction of radar in the Second World War, as well as by the extraordinary popularity of radio and television as the mass media of communication in the twentieth century. The increasing reliance in this century on techniques of mass production has encouraged the development of yet another type of engineering specialisation in the shape of the production engineer, for whom the Institution of Production Engineers was set up in 1921 to provide a professional society. It received a charter in 1964 and in 1979 it had 18,663 members.[5] A more recent foundation, and one even more characteristic of the twentieth century, was that of the Institution of Nuclear Engineers in 1959: this had 1,864 members in 1979.[6]

Both the 1914 and the 1979 lists of engineering professional institutions contain an element of double and multiple membership, but this is not sufficient to detract from the generalisations which can be made from a comparison between the two: first, that the total of professional engineers in Britain has grown ninefold in little more than half a century; and second, that this increase has tended to be contained in larger institutions rather than in more institutions. In other words, the rate of institutional proliferation has fallen off significantly, despite the overall increase in the number of institutions listed from seventeen to twenty-five. In particular, the three largest institutions in 1914 have all grown considerably, varying from a sixfold to a twelvefold increase. The Institution of Electrical Engineers has leapt from a membership of 7,045 to 74,928, making it the largest professional organisation of British engineers in 1979. The Institution of Mechanical Engineers has moved up from 6,400 to 72,654. And the Institution of Civil Engineers, still the largest body in 1914, with 9,194 members, had risen to 59,832, leaving it well behind its main rivals for leadership of the profession in terms of size. Nor is size the only evidence of diminished standing for the Civils. From being the only institution in 1914 with the status of a Royal Charter, the Civils now share this distinction with all the large engineering organisations. This decline in the relative position of the Civils has further weakened its influence - already undermined by institutional proliferation in the nineteenth century - to act on behalf of the whole profession. A new work on the history of the Civils gives the title 'Towards unity' to its chapter on developments from 1960 to 1978, but it has to be said that this represents an aspiration and that it is not an accurate description of the institutional changes of the period.[7]

The figures which have been used for membership of the engineering institutions in 1979 were those gathered by the Finniston Committee in the course of its analysis for the Report on *Engineering our Future*, published in 1980.[8] This Committee, under the Chairmanship of Sir Montague Finniston, FRS, had been invited to review the requirements of British industry for professional and technician engineers, and to consider the role of the engineering institutions in relation to the education and qualification of engineers. It would go beyond the limits of the present study to enter into a full account of the findings of the Finniston Committee, but a few comments are appropriate. For one thing, it is relevant to draw attention to the anxiety which produced the enquiry, because there is good reason to believe that contemporary apprehensions about the engineering profession and the role of the engineer in society have their roots in the nineteenth century. Most particularly, the widespread modern complaint amongst engineers that they do not receive the recognition which they deserve, which is strongly reflected in the Finniston Report, is related to the institutional proliferation of the second half of the nineteenth century. Although there were good reasons for this proliferation at the time, especially in the rapid increase in the range of highly specialised skills which members of the engineering profession were required to possess, in retrospect the disadvantages have come to outweigh the advantages of separate organisations. The pay-off for over-indulgence in specialist institutions has been a loss of professional coherence, in the sense that there is no one voice which is able to speak with authority for the whole profession. The ease with which proliferation was permitted and even encouraged, in the interests of maintaining the collective intimacy of every specialised sub-section of the profession, has now come to be widely regretted, as can be seen from the frequency with which it has recurred as a theme in presidential addresses to institutions since 1918. But the momentum of institutional existence is not easily reversed, and there is no simple course back to professional unity.[9]

Part of the problem of professional engineers in Britain is that they have come to feel at a disadvantage in comparison with engineers in other countries. It is, of course, true that the profession has developed differently in other countries, and that the pattern of organisation in Britain has some unique characteristics. But the extent of any consequent disadvantage is probably exaggerated. For one thing, it should be observed that the figures for membership of the professional institutions which we have just discussed point to a very healthy growth and suggest that the profession is essentially sound. Britain still possesses one of the largest communities of engineers in the world, both in absolute terms and in relation to the total population, so that in terms of numbers alone British engineers have no cause for serious concern.[10] And secondly, the success of the British pattern of professional organisation has been amply demonstrated by the readiness with which it has been imitated. We have seen how the Institution of Civil Engineers became a model for institutional organisation in Britain, the founders of new specialist associations going back again and again to the precedent established by the Civils in 1818 and the subsequent formative decades. In fact, however, the model has been even more widely applied, as engineers in other countries have sought to create a viable pattern of professional organisation for themselves.

This has been most apparent in the English-speaking countries. In the USA, moves were being made by the engineers towards institutional organisation from the 1830s. It was, to begin with, 'a sporadic and sometimes frustrating process',[11] with several hopeful starts stimulated by the activities of the railway engineers across the continent but then largely frustrated by the distances involved and the widely scattered distribution of engineering practitioners. The first permanent engineering organisation in America was the Boston Society of Civil Engineers, founded in 1848 with James Laurie as its first President. Laurie went on to become President of the American Society of Civil Engineers and Architects when this was founded in 1852, but with the increasing tension between the Northern and the Southern states which led to the Civil War in 1861-65, this lapsed into a condition described by its historian as 'suspended animation', from which it revived in 1867, with Laurie still presiding. The title was modified in 1868 by the omission of the words 'and Architects', because the American Institute of Architects had been formed as a separate body in 1857. Thenceforward, the ASCE grew surely and soundly. It had an international dimension from the start through its category of 'Corresponding Members', and it took positive measures to reinforce international contacts. The use of the Institution of Civil Engineers as a model had been acknowledged from the beginning, and formal links were established with the British institution in 1882. Various American delegations were entertained in Britain by the Civils in the following decades, and in 1904 the ASCE was host to a visiting party from Britain.[12]

By this time there had been some institutional proliferation amongst American engineers, but the process was not pursued so thoroughly as in Britain. The bulk of American professional engineers have been members of one of the four 'founder' societies which are, in addition to the ASCE, the American Institution of Mining Engineers (AIME), established in 1871; the American Society of Mechanical Engineers (ASME), established in 1880; and the American Institution of Electrical Engineers (AIEE), established in 1884. All of these adopted the type of organisation pioneered by the Civils, but similarities between the American and British situations go beyond institutional comparisons. The controversy among American mechanical engineers between the exponents of a 'shop culture' produced by the traditional machine-shops which bred a self-perpetuating elite for whom professional recognition meant a guarantee of gentlemanly status, and the advocates of a 'school culture' emphasising the importance of theoretical training to prepare engineers for salaried service in the new bureaucratic corporations, bore a striking resemblance to one of the recurrent tensions within the British engineering community in the closing decades of the nineteenth century.[13] The visit of a boat-load of American engineers to Britain in 1889, on their way to the Paris Exhibition, provided a pretext for serious reflection on the international interests and responsibilities of engineers, as well as for considerable festivity and social intercourse.[14]

A similar process of institutional replication had occurred in Canada, where there had been strong pressure in the second half of the nineteenth century from the rival cultural traditions of the British Motherland on the one hand, and the dynamic neighbour to the south on the other. Canadians preferred to adopt the more useful aspects of both traditions, while maintaining their independence of both. In relation to engineering, this involved a readiness to accept railway engineers from both Britain and the USA, while seeking to develop a distinctively Canadian style of railway construction. The Canadian government chose the 5ft 6ins gauge as the standard for its railway system, which led to difficulties in junctions with American lines built on the 4ft 8.5 ins gauge, and thus served to distinguish the Canadian network from that to the south. Of even greater significance for the development of a distinct Canadian engineering tradition was the formation of the Canadian Society of Civil Engineers in 1887. Like the ASCE and other American institutions, it was modelled on the Civils. But the CSCE acquired characteristics of its own, such as the emphasis which it came to place on legislative protection for Canadian engineers as a 'close corporation' against outside competition. This policy was never completely successful, and Canada continued to benefit from British and American engineering expertise.[15]

In Australia, despite considerable engineering enterprise in road building, railway construction, mining activities, and urban services, the great size of the continent and the

consequent dispersal of the engineers hindered the formation of permanent organisations. But the Engineering Association of New South Wales was established in 1870, on
the initiative of James Laing, and achieved its Act of Incorporation in 1884. Thereafter,
successful organisations were formed in Victoria and Queensland, and university engineering societies were established in 1889 at Melbourne, where the leading part was played
by Professor W. C. Kernot, and in 1895 at Sydney, where the chief spokesman was Professor W. H. Warren. Warren went on to become the first President of the Institution of
Engineers, Australia, when this was established in 1919, drawing on the pre-existing
'foundation societies' in the various states and universities for its initial membership of
over 1,000.[16] Institutional proliferation in Australia had been territorial rather than related to professional specialisation, but in achieving an almost complete integration of
their associations Australian engineers secured a professional unity which had been forfeited in Britain. Recent research has demonstrated that this achievement was reinforced
rather than impaired by the foundation of the Association of Professional Engineers in
Australia in 1946, because the APEA acted as a trade union in promoting the interests
of engineers in industrial relations and thus complemented the functions of the IEAust.
The prestige of the latter had been enhanced by the acquisition of a Royal Charter in
1938, and while it continued to perform the traditional functions of an engineering institution on the model of the Civils, the APEA concentrated on the material well-being
of Australian engineers.[17] Even though they managed to improve on the British model,
Australian engineering institutions thus remained very conscious of their British precedents.

The model provided by the British Institution of Civil Engineers was also reproduced
in non-English-speaking countries. Switzerland led the way with the formation of the So-
ciéte Suisse des Ingénieurs et des Architects in 1837. In the early stages of institutionalisation, engineers and architects frequently combined forces, although less so in Britain
than elsewhere because both architects and surveyors had made independent progress
towards professional status by the time that the Institution of Civil Engineers was established in 1818.[18] In the late 1840s, coinciding with the formation of the Institution of
Mechanical Engineers in Britain, three important institutions of the British type were established on the European mainland. These were the Koninlijk Institut van Ingenieurs
in the Netherlands in 1847, the Société des Ingénieurs Civil de France in 1848, and the
Osterreichischer Ingenieur und Architektenverein in Austria also in 1848. A few years
later, the same model was used for the Verein Deutscher Ingenieure which was formed
in Germany in 1856, having developed from an academic association of engineers
founded at the Royal Prussian College of Technology in Berlin in 1846.[19]

The Institution of Civil Engineers exchanged publications with several similar bodies
in the second half of the nineteenth century. These included institutions in Russia (the
Imperial Institution of Ways and Communications in St Petersburg, in 1861), in Portugal (the Society of Civil Engineers in Portugal, in 1872), and in Hungary (the Society of
Hungarian Engineers and Architects, which was distinct from the Austrian Society, in
1873). It maintained close relations with Indian engineers, many of whom during the
period of the Raj were its own members, and promoted local organisations such as the
Bombay Mechanics Institute, which received publications from the Civils in 1866. While

none of the overseas associations mentioned here were simply replicas of the Civils, the British model was acknowledged as the *primus inter pares*, and its advice was frequently sought on organisational details.[20]

Despite the facility with which the pattern of institutional organisation devised by the Civils was copied elsewhere, few continental institutions achieved the authority exercised in Britain by the Civils, mainly because alternative procedures for engineering training and professional qualifications had already been established, usually by the state. France, in particular, had developed an elaborate system of technological education on the basis of the *Grandes Écoles*, of which the École Polytechnique, founded in 1794, in the aftermath of the Revolution, was the outstanding example. Young men were selected for this programme of rigorous instruction by competitive examination, and on completion of the process they enjoyed an esteem which gave them an *entré* into the higher echelons of government service. There is nothing equivalent to this in the British system, and the Finniston Committee recently recorded its impression that the possessors of the prized and state-awarded title of 'Diplome d'Ingénieur' in France 'saw themselves much less as members of a profession than as members of an administrative elite'.[21] But other European countries have developed systems which are more closely akin to that of France than to Britain. The Federal Republic of Germany, for instance, has inherited the Prussian pattern of technological education based upon a broad secondary school curriculum and intensive specialised training in *Technische Hochschule*. This leads to the award of a 'Dipl Ing' qualification which is a *de facto* form of state registration of engineers. The Scandinavian countries have tended to follow the German model in this respect, with such distinguished institutions as the Royal Tekniska Högskola in Stockholm, and Chalmers Tekniska Högskola in Gothenburg, both established in the early nineteenth century, providing a supply of well trained engineers with qualifications recognised by the state.[22] It is arguable that the more home-spun procedures for the 'formation' of engineers practised in Britain in the late eighteenth and early nineteenth centuries - procedures of which the mutual improvement provided by institutions modelled on the Civils were an integral part - were more appropriate to the conditions of a nation in the early stages of industrialisation than were the more formal procedures which developed on the continent. Certainly, in the early years of rapid industrialisation, down to and including the great vogue for railway construction in the middle of the nineteenth century, it was British engineering which possessed the vital know-how and, as we have seen, British engineers who were in demand in Europe and throughout the industrialising world. It is possible that the British engineer in this period, with his more varied pattern of training and with the encouragement of his professional institution to keep abreast of recent developments in his field, was more flexible and adaptable than his continental counterpart. But in the second half of the nineteenth century the theoretical competence of the continental engineer gave him an advantage in the new fields of 'scientific' engineering, which was conceded by the steady advance of the component of theoretical instruction in the formation of British engineers, through university courses and alternative extra-mural training.

What was not so readily conceded in Britain, however, was the role of the profession in defining the procedures whereby engineers acquired their qualifications, and here, perhaps, the role of the professional institutions has been one which has caused confusion and loss of focus. While continental engineers have accepted some form of state registration, those in Britain have relied upon a less precise form of recognition which has had the effect of fogging the fundamental issues about the definition of engineering, causing tedious recriminations and, in the end, inaction. Combined with the refusal of the professional institutions to get involved in any form of labour relationships for their members, and their reluctance to promote an alternative form of industrial relations procedure (here the example of Australian engineering is instructive, as we have noted), and the lack of any effective alternative to the Civils as the mouthpiece for the whole profession - despite the efforts of the Council of Engineering Institutions since 1965 and, more recently, the Engineering Council, to perform this function - this development has had the effect of confirming the contemporary British engineering profession in a posture which resembles that of an inferiority complex. It is uncertain about its scope and its status; it is uneasy about its relations with other professions and with society at large; and it senses that it is losing out in competition with other traditions of engineering formation. And the confused response to the Finniston Report has done nothing to relieve these misgivings.

Whatever the sources of the prevalent inferiority complex which has afflicted the British engineering profession in the twentieth century, there are grounds for believing that it is at least partially misconceived. Even though their proliferation has been a cause for embarrassment to the profession, the engineering institutions represent a remarkable British achievement as organisations which stimulate professional self-awareness and continuing education amongst their members. The service of engineers to modern society has been substantial, and deserves acknowledgment. But in their wider social relationships, British engineers have become aware of situations in which they feel themselves, rightly or wrongly, to be operating at a disadvantage in comparison with other professional people, or with engineers of similar rank in other countries. The anxiety evoked by this awareness may be regarded as having derived from three overlapping sets of problems, concerned with identity, training and participation. In bringing our study of the British engineering profession to a conclusion, it will be appropriate to review the nature of these problems, and the possible responses to them.

The problem of identity arises in large part from the centrifugal tendency generated by the multiplicity of specialised institutions, whereby no single organisation or voice can represent the whole profession. It has become a broader issue because this deficiency calls in question the limits of the profession itself, as some of the newer institutions have struggled for recognition within the charmed circle of professional status. The continued openness of the profession to an inflow of individuals and groups from the ranks of technicians, artisans, foremen and other sub-professional levels, has been encouraged by the imprecision of the term 'engineer' in English usage. A recent sociological review has observed that, 'Measured in status terms, British engineering does not, as an occupation, rank very highly,even though some individual engineers do',[23] and it is difficult to deny the strength of this generalisation, even though, as we argued at the outset, it is a fruitless exercise to enter into an elaborate discussion of the nature of professionalism, and of whether or not engineering should be regarded as a profession. For the purposes of this study, the problem of definition has been overcome by adopting the term 'professional engineer' in all sensitive circumstances.Unfortunately for the engineering profession, no unanimity about its limits has existed since the institutional fragmentation of the profession in the second half of the nineteenth century, so that it has been impossible for British engineers to achieve complete self-assurance on the question of their identity. Commenting on the decline in the relative prestige of the Civils, J. O. Marsh has recently remarked that: 'Whatever the relevance of institution membership may have been in the past, the institutions are now not important in the eyes of employers'.[24] In this he echoed one of the more sombre findings of the Finniston Report: 'Employers... almost unanimously expressed their indifference to the benefits of Institution membership for the engineers they employ and were openly sceptical of the ability or authority of the Institutions to regulate the practice of engineering'.[25] In short, it seems that the various attempts to fill the vacuum left by the relative decline of the Civils since 1914 - the Engineering Joint Council, set up in 1924; the Engineering Institutions Joint Council, set up in 1962; and the Council for Engineering Institutions, chartered in 1965 - have all failed to be effective in arresting the loss of prestige of the engineering institutions. It is possible that the Engineering Council - set up in 1983 as a result of the Finniston Report, although with only a fraction of the powers required for the 'Engineering Authority' envisaged by that Report - may go some way towards overcoming the prevalent institutional anarchy by providing acceptable professional leadership for British engineers, but it is still too early to discern whether or not it will be sufficiently strong to do so. Until such a centripetal force is asserted, however, the centrifugal legacy of nineteenth century institutional proliferation will make the problem of engineering identity insoluble, and the profession will remain bound by the historical bonds which it has forged for itself.

Training constitutes the second problem area bearing on the sense of inferiority under which British professional engineers have tended to labour in the twentieth century, because they have allowed themselves to be convinced that in some respects they are not trained as well as their counterparts in other countries. There have been times in the history of British engineering, as we have seen in discussing the development of education and training, when there have been grounds for this apprehension.[26] In the second half of the nineteenth century, in particular, British engineering fell behind that in continental Europe and North America in failing to recognise the importance of formal theoretical training in the instruction of its recruits. But even in this period, British backwardness was less conspicuous than the advocates of technical education have represented it. The British institutions were not being simply blind, wilful, or self-interested in avoiding recognition of formal engineering education. The fact is that they had inherited a method of instruction - the traditional system of pupilage - which was well-tried and successful: in the light of the prestige enjoyed by engineering in Victorian Britain, it 'got results'. All the leaders of the profession were products of this system, so that it was difficult to convince such an impressive range of self-made men that it was capable of improvement. Of course, they were obliged in time to accept the logic of events, and especially that derived from the new theoretical skills in thermodynamics, electrical science, and chemical engineering, which were not easily assimilated in the course of pupilage with masters who had not acquired these skills themselves. So through the work of such

pioneers in academic engineering as Rankine, Fleeming Jenkin and Ewing, the universities offered an ingenious compromise: they provided courses in engineering theory
which were related to experimental lessons in laboratories and to practical experience,
and these became a prerequisite for many specialised forms of engineering, thus supplying theoretical skills without undermining the traditional insistence on work experience.

Equipped with this facility, the British engineering profession would have been justified in shaking off any sense of inferiority regarding its training procedures, for it had
achieved a pattern of engineering education which made the best of all worlds and which
was, indeed, as good as any obtainable elsewhere. The persistence of a sense of disadvantage in this respect is thus explicable in terms of psychological or habitual factors
rather than anything more substantial, although not necessarily less real on that account.
Because of the tardiness of the institutions in accepting the innovation, not all would-be
engineers were obliged to follow a training which amalgamated university instruction
with practical experience, and this led to confusion between different styles of engineering. And problems of access to the most appropriate forms of training are perennial, as
the nature of engineering is constantly changing, requiring a constant flexibility on the
part of educational institutions. But these reservations should not throw doubt on the
basic quality of the British training process, which is not justified by an objective comparative judgment. Once again, the British engineering profession is a prisoner of its own
history, in being unable to shake off the stigma of obsolete criticisms.

A third problem area in the social relationships of professional engineers in twentieth
century Britain is that of their participation in social activities other than those directly
connected to engineering. We have already had cause to observe that engineers were reluctant, in the nineteenth century, to adopt explicit or unconformist political and religious positions, although there were a few distinguished exceptions.[27] This tendency has
been even stronger in the twentieth century, so that British engineers have come to be
regarded as somewhat negative in all relationships beyond those with a specifically professional character. A debate in the House of Commons in March 1988 demonstrated
the tendency in several ways. For one thing, it was a Friday morning debate, thinly attended by a handful of government supporters and a single opposition member. The debate was on a motion urging that consideration be given to improving engineering
performance in British industries, and it was the first such debate since the discussion on
the Finniston Report in December 1980. Virtually all the speakers were impressed by
the importance of engineering in the national economy, but they deplored the reluctance
of young people to enter the profession, they criticised British firms for not promoting
more engineers to senior managerial positions, and they regretted that there were so few
engineers in the House. One speaker observed that, although there were sixty-eight barristers, thirty solicitors, twenty accountants, and six doctors in the House in 1988, there
were only two mechanical engineers.[28] Another recent study by Glover and Kelly has
drawn attention to the lack of engineers amongst the upper echelons of British industrial
management.[29] This is potentially one of the most serious failures of participation by
British engineers, so far as economic vitality is concerned.

The generalisation which should be drawn from this and other studies, while it may be unfair to a few individuals, is that British engineers *en masse* have become identified with a traditional indifference to non-professional commitments. This tradition is probably rooted in the engineering temperament or, more accurately, the temperament of men who are attracted to engineering, which tends to be dominated by a strong sense of the physical and tactile, being at ease with drawing and model-making, and being uncomfortable with more speculative and theoretical concepts. As Brooke Hindle has argued, such men prefer action to argument: they are doers rather than talkers - people with a highly-developed sense of design, whose images tend to be three-dimensional rather than abstract.[30] Again, it is easy to be less than just to particular engineers in such generalisations, but there is a widespread consensus that such a distinctive engineering temperament exists, and that it has tended to discourage engineers from commitments apart from matters which engross their immediate professional attention. Thus they have acquired a reputation for non-participation. They may behave as excellent family men and law-abiding citizens, but their hard work and indifference to other commitments has meant that they have rarely been known for their social, political or religious activities, as industrial entrepreneurs and managers, as patrons of the arts and sciences, or as speculators about man and society. It is a matter for conjecture why these qualities should seem more characteristic of British engineers than those, say, of America, where engineers have frequently exercised industrial and social power, but in this as in other respects the reason is probably related to the different social environments in which the historical traditions have matured.

There is a further dimension to the problem of the social participation of engineers which relates to our discussion of deindustrialisation in the previous chapter.[31] It could be argued that the withdrawal of some successful engineers into concern for their newly-acquired estates and enjoyment of their magnified social status, which became a well-attested phenomenon of the late nineteenth century, exemplifies the thesis of Martin Wiener that there was a decline in support for the 'industrial spirit' in Britain after 1850. In so far as engineers participated in this withdrawal of commitment it is possible that it contributed to the decline in the self-confidence of the profession. It seems more likely, however, that engineers, as natural conformists in social affairs, adopted such aspirations because they were the accepted social goals, and that these goals were particularly attractive to engineers because they did not involve any unusual or unorthodox affiliations. Either way the relative withdrawal is interpreted, the attitude has been one which has accentuated the engineers' sense of social isolation, because the effect of avoiding participation in social and political activities has deprived the profession of spokesmen who might otherwise have provided a powerful case for them. This tendency was confirmed by the self-immolation of several senior Victorian engineers in their estates and country pursuits. Here, as elsewhere, the history of the engineering profession provides some clues to the problems of contemporary engineers.

There is thus substantial evidence that the problems of the British engineering profession at the end of the twentieth century are in large measure the result of the shape which the profession assumed during the formative decades of its evolution in the nineteenth century. Although these problems are real enough, however, it is worth recalling in conclusion that the history of the British engineering profession in the eighteenth and nineteenth centuries was a most remarkable success story, and that in several outstanding respects the engineering community of today continues to derive benefits from these historical roots. The success was one of accommodating the profession to a rapid expansion in the demand for engineering skills, by supplying a steady increase in the numbers of men able to perform increasingly complex and specialised tasks in a flourishing industrial society. The opportunities were thrust upon the profession by the circumstances of rapid industrialisation, but the engineering community deserves credit for the sound leadership and sober instincts with which it responded to these escalating demands.

The most important instrument of this response was the engineering professional institution, as an organisation of mutual instruction which was able to maintain the standards of entry into the profession, to encourage members to improve their skills in order to climb through a hierarchy of grades, to promote specific training programmes for engineers, and to monitor the conduct of their members, applying disciplinary measures where necessary. This exemplary organisation was first devised in the Institution of Civil Engineers in 1818, although it had significant precursors, and it took a couple of decades to work out the full implications of the model. Thereafter, the pattern was adopted repeatedly in Britain and elsewhere, whenever and wherever a group of specialist engineers sought the professional identity of a distinct organisation. We have examined the anomalies and weaknesses which have derived from the proliferation of such institutions in Britain. But proliferation has been a remarkable tribute to the success of this type of organisation, and it is not too much to say that the creation and replication of the engineering institution on the model of the Civils, eventually performing all the functions of a mature professional organisation, was one of the outstanding administrative achievements of any professional group in the nineteenth century. It made possible a relatively smooth response to the mounting engineering demands of an industrialising society. And it gave its members a degree of mutual support which ensured their loyalty and assured a large number of otherwise anonymous engineers of some identity as part of a creative association.

Notes

1. The difficulties of this interpretation are discussed above in Chapter Ten. For a recent stimulating but rather tendentious account, see Martin J. Wiener, *English Culture and the Decline of the Industrial Spirit 1850-1980*, Cambridge, 1981.

2. The figures for 1979 are given in tabular form in the *Finniston Report* (see below, note 8), p. 126.

3. Sir Arthur Duckham, KCB, Presidential Address, *Trans. Inst. Chem. Engs.*, 2, 1924, 14. See also J. B. Brennan, *The First Fifty Years - A history of the Institution of Chemical Engineers 1922-1972*, typescript produced by the Institution for Mr Brennan, General Secretary 1950-1969.

4. *A twentieth century professional institution - The story of the Brit. I.R.E.*, Institution of Electronic and Radio Engineers, London, 1960.

5. 'The Institution of Production Engineers - Brief History' in *Jnl. Inst. Prod. Engs*, 26, 1947, 145-6.

6. Figure as cited in *Finniston Report*, p.126. The Institution publishes *Nuclear Energy*: see 'The Institution and the future', May/June 1968, p. 85.

7. Garth Watson, *The Civils - The story of the Institution of Civil Engineers*, Thomas Telford, London, 1988: Chapter Five is entitled 'Towards unity: 1960-1978'.

8. Sir Montague Finniston, FRS (Chairman), *Engineering our Future - Report of the Committee of Inquiry into the Engineering Profession*, HMSO, London, January 1980, Cmnd.7794.

9. The sense of regret about institutional fragmentation was already becoming apparent before 1914 - see above, Chapter Six. But it subsequently became a familiar refrain in the presidential addresses to the various institutions, as reported regularly in their *Transactions*.

10. It is interesting to note that Göran Ahlström, who is not inclined to be over-sympathetic towards the British tradition of engineering history, calculated that there were about 60,000 'highly qualified engineers' in Germany in 1914, and slightly over 40,000 in France: *Engineers and Industrial Growth*, Croom Helm, London and Canberra, 1982, p. 13. The figure of just over 40,000 members of British engineering institutions at this date bears comparison with these statistics and demonstrates that, as a proportion of total population, the numbers of British engineers were much the same as in other industrialised European countries in 1914.

11. William H. Wisely, *The American Civil Engineer 1852-1974 - The history traditions and development of the American Society of Civil Engineers, founded 1852*, New York, 1974, p.1.

12. See Watson, *op. cit.*, 239-40. The 1904 visit coincided with the St. Louis Exhibition, and members went on to visit the Canadian Society of Civil Engineers on their way home.

13. Monte A. Calvert, *The Mechanical Engineer in America, 1830-1910 - Professional Cultures in Conflict*, Baltimore, 1967.

14. Bruce Sinclair, *The Centennial History of the American Society of Mechanical Engineers, 1880-1980*, Toronto, 1980, 36-38. See also - Raymond H. Merritt, *Engineering in American Society 1850-1875*, Kentucky, 1969.

15. Bruce Sinclair, 'Canadian Technology: British Traditions and American Influences', in *Technology and Culture*, 20, 1, January 1979, 108-123.

16. A. H. Corbett, *The Institution of Engineers, Australia*, Sydney, 1973. See also - R. A. Buchanan, 'The British Contribution to Australian Engineering' in *Historical Studies*, University of Melbourne, 1983, 401-419.

17. B. A. Lloyd, *In Search of Identity, Engineering in Australia 1788-1988*, unpublished PhD thesis, University of Melbourne, 1988. Lloyd considers that 'The degree of control achieved through industrial relations superseded the necessity for the pursuit of government sanction through statutory regulation' - p.173.

18. The Royal Institute of British Architects dates from 1834 - see Barrington Kaye, *The Development of the Architectural Profession in Britain*, London, 1960. For the Surveyors, see - F. M. L. Thompson, *Chartered Surveyors - The growth of a profession*, London, Routledge, 1968: the Institution of Surveyors was founded in 1868, but Thompson demonstrates that it had been preceded by several preliminary organisations.

19. R. A. Buchanan, 'Engineering in the International Community', in *Proc. 16th International Congress of the History of Science*, pt.B, Symposia, 30-35, Bucharest, 1981.

20. Details of the exchanges between the Civils and similar bodies are given in the *Proc. Inst. Civil Engs.* in the Annual Reports for the years mentioned.

21. *Finniston Report*, p. 217. See also Terry Shinn, 'From "corps" to "profession": the emergence and definition of industrial engineering in modern France' in Robert Fox and George Weisz (eds), *The organisation of science and technology in France 1808-1914*, Cambridge University Press, 1980.

22. *Finniston Report*, Appendix E, 199-237, 'Visits Overseas': see also Ahlström, *op. cit.*, for comparisons in engineering history between various European countries.

23. Ian A. Glover and Michael P. Kelly, *Engineers in Britain - A Sociological Study of the Engineering Dimension*, Allen and Unwin, London, 1987, p. 24.

24. J. O. Marsh, 'The engineering institutions and the public recognition of British engineers', in *The International Journal of Mechanical Engineering Education*, 16, 2, April 1988, 119-127, quoting from p. 126.

25. *Finniston Report*, p. 129.

26. See above, Chapter Nine.

27. See above, Chapter Ten.

28. *Hansard Parliamentary Reports*, House of Commons, 11 March 1988, cols.763-831, referring to the speech by Sir Barney Hayhoe, col.772.

29. Glover and Kelly, *op. cit.*, see especially Chapter 11, 'Engineering work: the division of labour' - 'Engineers are more concerned about whether things work, rather than why they work' - p. 209, and 'Qualified British engineers are not part of the upper-class elite as a group...' - p. 216.

30. Brooke Hindle, *Emulation and Invention*, New York, 1981: the point is lucidly made in Chapter 6, 'The Contriving Mind'.

31. See above, Chapter Ten.

Bibliography

I: INSTITUTIONAL SOURCES

1. British National Institutions

Parliamentary Papers

Hansard Parliamentary Proceedings.

'Report of the Commission on the Application of Iron to Railway Structures', *P.P. 1849* c.1123, xxix.

'Report of the Court of Inquiry... upon the circumstances attending the fall of a portion of the Tay Bridge on the 28th December 1879', *P.P. 1880* c.2616, xxxix.

Institution of Automobile Engineers
Fd.1906

Proceedings

Institution of Chemical Engineers
Fd.1922

Transactions from 1923:Gibb, A, 'The coordination of engineering institutions and societies' 7, 1929, 13-16.

Institution of Civil Engineers
Fd.1818 Royal Charter 1828

Transactions

Minutes of Proceedings: from 1837 - vol.1 covered 1837-1841, vol.2 covered 1842 and 1843, vol.3 covered 1844, and annual thereafter.

A brief history... with an account of the Charter Centenary Celebration, June 1928 London 1928.

The education and status of civil engineers, in the United Kingdom and in Foreign Countries London 1870.

Engineering Education in the British Dominions compiled from official sources, with the regulations of the Institution of Civil Engineers as to the admission of students London 1891.

EUSEC Conference on Engineering Education London 1953.

Dennis, T. L. (ed.), *Engineering societies in the life of a country*, London, 1968.

Watson, G., *The Civils - The story of the Institution of Civil Engineers*, Thomas Telford, London, 1988.

Institution of Electrical Engineers
Fd.1871 as 'Society of Telegraph Engineers'; assumed new name 1888, Incorporated 1883, Royal Charter 1921.
Journal: from 1873
Appleyard, R., *The History of the Institution of Electrical Engineers (1871-1931)* London, 1939.
Reader, W. J., *A History of the Institution of Electrical Engineers 1871-1971*, Peter Peregrinus, London, 1987.

Institution of Electronic and Radio Engineers
Fd.1925, Charter 1961.
Journal: from 1925.
A Twentieth Century Professional Institution - The Story of the Brit.I.R.E., London, 1960.

Institution of Gas Engineers
Fd.1863 but underwent several changes: incorporated 1890 and again in 1902: acquired Charter 1929
Transactions: annually from 1903
Braunholtz, W. T. K., *The First Hundred Years 1863-1963*, London, 1963.
Haffner, A. E., 'Centenary Presidential Address', *Jnl.Inst.Gas Engs.* 3, 1963.

Institution of Heating and Ventilating Engineers
Fd.1897
Proceedings: annually from 1899

Institution of Locomotive Engineers
Fd.1911
Journal: from 1911
Holcroft, H., 'The First Forty Years', 50, 1960-61, 662-682.

Institute of Marine Engineers
Fd.1889
Transactions: from 1889
Curling, B. C., *History of the Institute of Marine Engineers*, London 1961.

Institution of Mechanical Engineers
Fd.1847, Royal Charter, 1930.
Proceedings: from 1847
Parsons, R. H., *A History of the Institution of Mechanical Engineers, 1847-1947* London 1947.
Rolt, L. T. C., *The Mechanicals - progress of a profession*, London 1967.
Mowat, M., 'British engineering societies and their aims' - *Procs*, 137, 1937, 333-343.

Institution of Mining Engineers (Federated)
Fd.1889 by federation of regional associations: Charter 1918
Transactions: annual from 1890

Institution of Mining and Metallurgy
Fd.1892, Charter 1915
Transactions: annual from 1892

Institution of Municipal Engineers
Fd.1873 as Association of Municipal and Sanitary Engineers and Surveyors: Charter 1890

Proceedings: from 1873.

Institution of Naval Architects
Fd.1860: Charter 1910: 'Royal' 1959
Transactions: from 1860
Barnaby, K. C., *The Institution of Naval Architects 1860-1960*, London, 1960.
White, Sir William, *The History of the Institution of Naval Architects and of Scientific Education in Naval Architecture*, London, 1911.

Institution of Production Engineers
Fd.1921. Charter 1964.
Journal: from 1921
'Brief History', *Jnl.Inst.Prod.Engs.*, vol.26, 1947, pp.145-7.

Institution of Royal Engineers
Fd.1875 as 'The Royal Engineers' Institute': Charter 1923
Journal
Charter, Bye-Laws and Rules Chatham 1928
Professional Papers of the Corps of Royal Engineers, vol.1, second ed., 1844
Porter, W., *History of the Corps of Royal Engineers* London 1889
The Royal Engineers: Being remarks on their duties... London 1862
The Royal Engineer Department: its work and the Estimates, by 'Argus', London 1862.

Institute of Sanitary Engineers
Fd.1895: incorporated as 'Institution' in 1916; became the Institution of Public Health Engineers in 1955
Journal: from 1896

Institution of Structural Engineers
Fd.1908 as the 'Concrete Institute'; incorporated 1909; adopted present title 1922; Royal Charter 1934
Vaughan, S., 'Presidential Address 1955', *The Structural Engineer*, 33, December 1955, 365-375
Hamilton, S. B., 'The history of the Institution of Structural Engineers', *The Structural Engineer*, July 1958, 16-21
Fox, C. F., 'The history and progress of the Institution of Structural Engineers', *The Structural Engineer*, 13, n.s., 1935.

Institution of Water Engineers
Fd.1896 as 'Association of Water Engineers'
Transactions: from 1896
Peters, N. J., 'Presidential Address 1936', *Trans.I.Water E.*, 41, 1936, 16-17.

Iron and Steel Institute
Fd.1869
Journal: from 1871

Royal Aeronautical Society
Fd. 1866
A Short History... published by the Society, 1966.

Royal Society
Fd.1660
Philosophical Transactions from 1665

Society of Civil Engineers (Smeatonians)
Minute Book 1771-92
Minute Books from 1793
Donkin, S. B., 'The Society of Civil Engineers (Smeatonians)', *Trans.Newcomen Society*, 17, 1936-7, 51-71
Wright, Esther C., 'The Early Smeatonians', *Trans.Newcomen Society*, 18, 1937-8, 101-110
Skempton, A. W. and Wright, Esther C., 'Early members of the Smeatonian Society of Civil Engineers', *Trans.Newcomen Society*, 44, 1967-8, 23-47.

Society of Engineers
Fd.1854 Incorporated 1910
Transactions: annual from 1861

2. British Regional Institutions

Belfast Association of Engineers
Marr, D. B., *Recollections and Memoirs of the Belfast Association of Engineers*, Belfast, 1967.

Birmingham Association of Mechanical Engineers
Fd.1889: originally 'Birmingham and District Foremen and Draughtsmen Mechanical Association', adopted later title 1891: no activities reported after 1942
Proceedings 1900-1940
Mackintosh, D. G., 'Presidential Address', *Proc.* 1928-29: [Birmingham City Library - L65.206 etc.]

Chesterfield and Derbys Institute of Mining and Mechanical Eng.
Fd.1871
Hinsley, F. B., 'A Centenary History....' in *The Mining Engineer*, 1976, 3 parts(May 1976, 493-500; June 1976, 561-564; July 1976, 623-627).

Glasgow and West of Scotland Association of Foremen Engineers and Draughtsmen
Fd.1898
Annual Reports [a few volumes in Mitchell Library Glasgow]

Institution of Civil Engineers of Ireland
Fd.1835, re-fd.1844
Transactions:
Griffith, Sir Richard, 'Inaugural Address', 6,1861,193-221
Hemans, G. W., 'Presidential Address', 5, 1860, 51-65
Mallet, R., 'Presidential Address', 8, 1868, 48-102
Mullins, M. B., 'Presidential Address', 6, 1861, 1-186
Also:
Cox, R. C., *Engineering Ireland 1778-1878*, Exhibition Catalogue, School of Engineering, TCD, Dublin, 1978; and Hughes, N. J., *Irish Engineering 1760-1960*, published by the Institution, 1982.

Institution of Engineers and Shipbuilders in Scotland
Fd.1857
Transactions since 1857.

Keighley Association of Engineers
Fd.1900
Programmes.

Leeds Association of Foremen Engineers and Draughtsmen
Fd.1865
Annual Reports since 1865.

Liverpool Engineering Society
Fd.1875
Transactions from 1881 to 1960.

Mills, W. E., 'Fifty years of the Liverpool Engineers Society', 47, 1926, 171-204.

Journal from 1960 to 1968.

[About thirty volumes and bundles were deposited in Liverpool City Archives and Local History Library in 1973 by Dr S. J. Kennett, the Senior Vice-President of the Society: they are classified under 620 ENG-Acc.2561]

Manchester Association of Engineers
Fd.1856
Transactions:

'Address of Mr.Thomas Ashbury, CE, on his being elected for the second time President...' 1884

Dean, A. C., *Some episodes in the Manchester Association of Engineers*, Manchester, 1938.

The Manchester Association of Engineers 1856-1956: A hundred years of engineering in Manchester, Manchester, 1956.

Manchester Institution of Engineers
Fd.1867
No activities recorded after 1868
Proceedings, 1867-8.

Miners' Association of Cornwall and Devonshire
Fd.1859: amalg. with Mining Institute of Cornwall to form 'Mining Association and Institute' 1885.
No activities after 1895.

Report of Annual Meeting: series 1866-1876

Proceedings (of the Institute) 1876-1883

Transactions (of the joint body) from 1885.

Also:

Transactions, Cornish Institute of Mining and Mechanical Engineers, 1, 1913.

North-East Coast Institution of Engineers and Shipbuilders
Fd.1884
Transactions from 1885:

Boyd, W., 'Inaugural Address' as President, 1, 1885.

Nicol, T. C., 'The Institution', 77, 1961,151-166.

North of England Institute of Mining Engineers
Fd.1852: 'and Mechanical' added 1866.
Council Minute Books from 1852.
Transactions, 1, 1852-3.
Centenary Brochure 1852-1952, Newcastle, 1952.

Sheffield Society of Engineers and Metallurgists
Fd.1894: no activities reported after 1949.
Rules and Lists of Members, 1914. Sheffield City Library: *Local Pamphlets*, no. 2.
Formed by amalgamation of Sheffield Soc. of Engs.(1888) and Sheffield Metallurgical Soc.(1890).

South Wales Institute of Engineers
Fd.1857: incorporated 1881.
Proceedings.
Centenary Brochure 1857-1957, Cardiff 1957.

3. Other Institutions

American Society of Civil Engineers
Fd.1857, re-fd.1867.
Wisely, W. H., *The American Civil Engineer 1852-1974*, New York, 1974.

American Society of Mechanical Engineers
Fd.1880
Calvert, M. A., *The Mechanical Engineer in America, 1830-1910: Professional Cultures in Conflict*, Johns
   Hopkins Press, Baltimore, 1967.
Sinclair, Bruce, *A Centennial History of the American Society of Mechanical Engineers*, ASME, Toronto, 1980.

Institution of Engineers, Australia
Fd.1919.
Corbett, A. H., *The Institution of Engineers Australia: A History of the First Fifty Years 1919-1969*, Angus and
   Robertson, Sydney, 1973.

II: ENGINEERING BIOGRAPHY

Abrams, John W. 'Sandford Fleeming in Canada', *Trans.Newcomen Soc.*, 49, 1977-78, 133-137.
Appleyard, Rollo, *Charles Parsons: His Life and Work*, Constable, London, 1933.
Aspin, C. and Chapman, S. D., *James Hargreaves and the Spinning Jenny*, Helmshore Local History Society
   1964.
Bailey, Frank, *The Life and Work of S. Z. de Ferranti DSc FRS 1864-1930*, Ferranti Ltd. Hollinwood Lancs,
   1931.
Baker, E.C., *Sir William Preece, FRS, Victorian Engineer Extraordinary*, Hutchinson, London, 1976.
Banks, A. G. and Schofield, R. B., *Brindley at Wet Earth Colliery: an Engineering Study*, David and Charles,
   Newton Abbot, 1968.
Beamish, Richard FRS, *Memoir of the Life of Sir Marc Isambard Brunel*, Longman, London, 1862.
Beckett, Derrick, *Brunel's Britain*, David and Charles, Newton Abbot, n.d. 1980?

Beckett, Derrick, *Stephenson's Britain*, David and Charles, Newton Abbot, 1984.

Bell, S. P. (compiled by), *A Biographical Index of British Engineers in the 19th Century*, Garland Publishing Inc., New York and London, 1975.

Bellwood, John and Jenkinson, David, *Gresley and Stanier: A Centenary Tribute*, National Railway Museum York and HMSO, 1976.

Bessemer, Henry, *Sir Henry Bessemer, FRS - An Autobiography*, London: Offices of 'Engineering', 1905.

Binnie, G. M., *Early Victorian Water Engineers*, Thomas Telford Ltd., London, 1981.

Bishop, R. E. D., 'Alexander Kennedy - the elegant innovator', *Trans.Newcomen Soc.*, 47, 1974-76, 1-8.

Booth, L. G. et.al., 'Thomas Tredgold (1788-1829) - Some aspects of his work', *Trans.Newcomen Soc.*, 51, 1979-80, 57-94.

Boucher, Cyril T. G., *James Brindley Engineer 1716-1772*, Goose, Norwich, 1968.

Boucher, Cyril T. G., *John Rennie 1761-1821: The Life and Work of a Great Engineer*, Manchester University Press, 1963.

Brunel, Isambard, *The Life of Isambard Kingdom Brunel - Civil Engineer*, Longman, London, 1870, reprinted David and Charles, Newton Abbot, 1970.

Brunel, I. K., *Private Letter Books*, 11 manuscript vols. in the Archives of Bristol University Library.

Brunton, John, *John Brunton's Book*, Cambridge University Press, 1939.

Chaloner, W. H., 'John Galloway (1804-1894), Engineer, of Manchester...', *Trans. Lancs.and Cheshire Antiq.Soc.*, 64, 1954.

Channell, David F., 'The harmony of theory and practice: the engineering science of W. J. M. Rankine', *Technology and Culture*, 23, 1, January 1982, 39-52.

Church, William Conant, *The Life of John Ericsson*: Vol. 1, London 1890; Vol. 2, London 1892.

Clark, E. F., *George Parker Bidder - The Calculating Boy*, KSL Publications, Bedford, 1983.

Clark, Ronald W., *Edison: the man who made the future*, MacDonald and Jane's, London, 1977.

Clements, Paul, *Marc Isambard Brunel*, Longman, London, 1970.

Conder, F. R. (Ed., Jack Simmons), *The Men Who Built Railways*, Thomas Telford Ltd., London, 1983.

Crompton, R. E., *Reminiscences*, Constable, London, 1928.

Davies, Hunter, *George Stephenson: A Biographical Study of the Father of Railways*, Weidenfeld and Nicolson, London, 1975.

Devey, Joseph, *The Life of Joseph Locke, Civil Engineer, MP, FRS, etc. etc.*, Richard Bentley,London, 1862.

Dickinson, H. W., *Matthew Boulton*, Cambridge University Press for Babcock and Wilcox Ltd., 1936. *James Watt: Craftsman and Engineer*, Cambridge University Press for Babcock and Wilcox Ltd., 1935.

Dickinson, H. W. and Jenkins, Rhys, *James Watt and the Steam Engine*, Oxford, 1927, reprinted Moorland, 1981.

Dickinson, H. W. and Titley, Arthur, *Richard Trevithick: The Engineer and the Man*, Cambridge University Press, 1934.

Dougan, David, *The Great Gun-maker: The Story of Lord Armstrong*, Frank Graham, Newcastle upon Tyne, 1970.

Ellis, C. Hamilton, *Twenty Locomotive Men*, Ian Allan Ltd., London 1958.

Emmerson, George S., *John Scott Russell: A Great Victorian Engineer and Naval Architect*, John Murray, London, 1977.

Ewing, Sir Alfred, *An Engineer's Outlook*, Methuen, London, 1933.

Ewing, A. W., *The Man of Room 40: The Life of Sir Alfred Ewing*, Hutchinson, London, 1939.

Fairbairn, William, *The Life of Sir William Fairbairn, Bart.*, edited and completed by William Pole, 1877, reprinted with introduction by A. E. Musson, David and Charles, 1970.

Fox, Sir Francis, *River, Road and Rail: some engineering reminiscences*, John Murray, London, 1904. *Sixty-three years of engineering - scientific and social work*, John Murray, London, 1924.

French, Gilbert J., *Life and Times of Samuel Crompton*, first edition 1859,second edition 1860, reprinted 1970 by Adams and Dart with Introduction by Stanley D. Chapman.

Froude, William, *The Papers of William Froude MA LLD FRS 1810-1879*, Institution of Naval Architects, London, 1955.

Gibb, Sir Alexander, *The Story of Telford: the Rise of Civil Engineering*, A. Maclehose and Co., London, 1935.

Hadfield, Charles, 'James Green as canal engineer', *Jnl.Transport Hist.*, 1, 1953-54, 44-56.

Hadfield, Charles and Skempton, A. W., *William Jessop, Engineer*, David and Charles, Newton Abbot, 1979.

Harris, L. E., *Vermuyden and the Fens: A study of Sir Cornelius Vermuyden and the Great Level*, Cleaver-Hume Press Ltd., London, 1953.

Harris, T. R., *Arthur Woolf: The Cornish Engineer 1766-1837*, Bradford Barton, Truro, 1966.

Harrison, Godfrey, *Alexander Gibb: The Story of an Engineer*, Geoffrey Bles, London, 1950.

Hay, Peter, *Brunel: his achievements in the transport revolution*, Osprey, Reading, 1973.

Jeaffreson, J. C., *The Life of Robert Stephenson, FRS*, Longman, London, 1864, 2 vols.

Kirby, Richard S., 'William Weston and his contribution to early American engineering', *Trans.Newcomen Soc.*, 16, 1955-56, 111-128.

Korthals-Altes, J., *Sir Cornelius Vermuyden: The lifework of a great Anglo-Dutchman in land-reclamation and drainage*, Williams and Norgate, London: Van Stockum, The Hague, 1925.

Lea, F. C., *Sir Joseph Whitworth: A pioneer of mechanical engineering*, Longman, for the British Council, London, 1946.

McDowell, D. M. and Jackson, J. D. (Eds), *Osborne Reynolds and Engineering Science Today*, Manchester University Press, 1970.

Mackay, Thomas, *The Life of Sir John Fowler Engineer Bart., KCMG, Etc.*, John Murray, London, 1900.

Mackenzie, Thomas B. (compiler), *Life of James Beaumont Neilson FRS*, The West of Scotland Iron and Steel Institute, Glasgow, 1929.

Mair, Craig, *A Star for Seamen: The Stevenson Family of Engineers*, John Murray, London, 1978.

Marshall, John, *A Biographical Dictionary of Railway Engineers*, David and Charles, Newton Abbot, 1978.

Mitchell, Joseph, *Reminiscences of my life in the Highlands*, originally pub. privately in 1883 in 2 vols., Reprinted David and Charles, Newton Abbot, 1971.

Moseley, Maboth, *Irascible Genius: A Life of Charles Babbage, Inventor*, Hutchinson, London, 1964.

Napier, David, *David Napier - Engineer - 1790-1869: An Autobiographical Sketch with Notes*, Maclehose, Glasgow, 1912.

Napier, James, *Life of Robert Napier*, Blackwood, Edinburgh and London, 1904.

Nasmyth, James, *James Nasmyth, Engineer: An Autobiography*, Ed. by Samuel Smiles, LL.D., John Murray, London, 1885.

Nelson, W., 'Josiah Hornblower and the first steam-engine in America', *Proc.New Jersey Hist.Society*, 2nd series, 7, 1883, 177-247.

Noble, Celia Brunel, *The Brunels - Father and Son*, Cobden-Sanderson, London, 1938.

Nock, O. S., *The Railway Engineers*, Batsford, London, 1955.

Peet, H., 'Thomas Steers - The Engineer of Liverpool's first dock - A Memoir', *Trans.Hist.Soc.of Lancs.and Cheshire*, 82, 1932, 163-242.

Penfold, Alastair (Ed.), *Thomas Telford: Engineer*, Thomas Telford Ltd., London, 1980.

Pudney, John, *Brunel and his World*, Thames and Hudson, London, 1974.

Pugsley, Sir Alfred (Ed.), *The Works of Isambard Kingdom Brunel: An Engineering Appreciation*, Institution of Civil Engineers, London, and University of Bristol, 1976.

Reader, W. J., *Macadam: The McAdam Family and the Turnpike Roads 1798-1861*, Heinemann, London, 1980.

Reid, Stuart J., *Sir Richard Tangye*, London, 1908.

Richardson, A. E., *Robert Mylne: Architect and Engineer 1733 to 1811*, Batsford, London, 1955.

Rickman, John (Ed.), *Life of Thomas Telford, Civil Engineer*, London, 1838.

Rogers, H. C. B., *G. J. Churchward: A Locomotive Biography*, Allen and Unwin, London, 1975.

Rolt, L. T. C., *Isambard Kingdom Brunel: A Biography*, Longman, London, 1957, Penguin 1975
    Thomas Telford, Longman, London, 1958. Penguin, 1979.
    George and Robert Stephenson: The Railway Revolution, Longman, London, 1960. Penguin, 1978.
    James Watt, Batsford, London, 1962.
    Thomas Newcomen: The Prehistory of the Steam Engine, David and Charles, Dawlish/Macdonald, London, 1963. Revised by J. S. Allen as *The Steam Engine of Thomas Newcomen*, Moorland, 1977.

Salmond, J. B., *Wade in Scotland*, Moray Press, Edinburgh and London, 1934.

Scott, E. Kilburn (ed.), *Matthew Murray, Pioneer Engineer, Records from 1765 to 1826*, Leeds, 1928.

Sharlin, Harold Issadore, *Lord Kelvin: The Dynamic Victorian*, Pennsylvania State University Press, 1979.

Skeat, W. O., *George Stephenson: The Engineer and His Letters*, Institution Mechanical Engineers, London, 1973.

Skempton, A. W. (Ed.), *John Smeaton FRS*, Thomas Telford Ltd., London, 1981.

Skempton, A. W., 'William Chapman (1749-1832), Civil Engineer', *Trans.Newcomen Soc.*, 46, 1973-74, 45-82.

Smeaton, J., *John Smeaton's Diary of his Journey to the Low Countries 1755*, Newcomen Society, Leamington Spa, 1938.
    Reports of the Late John Smeaton FRS, 3 vols., London 1812

Smiles, Samuel, *The Autobiography of Samuel Smiles, LL.D.*, Edited by Thomas Mackay, Author of the 'Life of Sir John Fowler' with 2 portraits, John Murray, London, 1905.
    Life of George Stephenson, London, first ed. 1857.
    Lives of Boulton and Watt, John Murray, London, 1865.
    Lives of the Engineers, 1862, reprinted David and Charles, 3 vols., Introduction by L. T. C. Rolt, 1968.
    Industrial Biography, London, 1882.
    Men of Invention and Industry, London, 1884.

Smith, Denis, 'The professional correspondence of John Smeaton...', *Trans.Newcomen Soc.*, 47, 1974-76, 179-189.

Stevenson, David, *Life of Robert Stevenson, Civil Engineer*, Black, Edinburgh etc. 1878.

Stevenson, Robert Louis, *Records of a Family of Engineers*, Chatto and Windus, London, 1912.
    Memoir of Fleeming Jenkin, Records of a Family of Engineers, Pub. together as Vol. 19 in the Tusitala Edition of the Works of R.L.S.: Heinemann, London, 1924.

Sutherland, Hugh B., *Rankine: His Life and Times*, Institution of Civil Engineers, London, 1973.

Tangye, Sir Richard, *'One and All': An Autobiography of Richard Tangye of the Cornwall Works, Birmingham*, London, 1889.
    Reminiscences of Travel in Australia, America, and Egypt, London and Birmingham, 1883.

Trevithick, Francis C. E., *Life of Richard Trevithick with an account of his inventions*, 2 vols. bound in one., London, E. and F.N.Spon., and New York, 1872.

Unwin, George, *Samuel Oldknow and the Arkwrights:, The Industrial Revolution at Stockport and Marple*, Manchester, 1924.

Vignoles, K. H., *Charles Blacker Vignoles: Romantic Engineer*, Cambridge University Press, 1982.

Vignoles, Olinthus J., *Life of Charles Blacker Vignoles FRS FRAS MRIA &c*, Longman, London, 1889.

Walker, E. G., *The Life and Work of William Crawthorne Unwin*, London, 1947.

Webster, N. W., *Joseph Locke: Railway Revolutionary*, George Allen and Unwin, London, 1970.

Wilson, Roger Burdett (Ed.), *Sir Daniel Gooch: Memoirs and Diary*, Transcribed from the original manuscript and edited with an Introduction and Notes..., David and Charles, Newton Abbot, 1972.

Whittle, Sir Frank, *Jet: the story of a pioneer*, first published by Fred Muller, 1953; Pan Books ed. 1957.

Young, Robert, *Timothy Hackworth and the Locomotive*, Shildon, Co. Durham, 1923, new ed. 1975.

III: GENERAL PUBLISHED AND UNPUBLISHED SOURCES

Agricola, Georgius, *De Re Metallica*, Basle, 1556: Dover ed. translated H. C. and L. H. Hoover, New York, 1950.

Ahlström, G., *Engineers and Industrial Growth*, Croom Helm, London and Canberra, 1982.

Argles, M., *South Kensington to Robbins: An account of English Technical and Scientific Education since 1851*, Longman, London, 1964.

Armytage, W. H. G., *Civic Universities*, Ernest Benn, London, 1955.

Armytage, W. H. G., *A Social History of Engineering*, Faber and Faber, London, 1961.

Ashby, Sir Eric, *Technology and the Academics*, Macmillan, London, 1959.

Ashton, T. S., *The Industrial Revolution 1760-1830*, Opus, Oxford, 1948.

Ashworth, W., *Economic History of England 1870-1939*, Methuen, London, 1960.

Ashworth, W., 'Economic Aspects of Late Victorian Naval Administration', *Econ. Hist. Review* second series 22, 3, 1969, 491-505.

Babbage, C., *Reflections on the Decline of Science in England*, London, 1830, and Irish University Press, 1971.

Babbage, C., *On the Economy of Machinery and Manufactures*, Charles Knight, London, third edition (enlarged) 1832, first and second editions 1832.

Barker,T. C., 'The Sankey Navigation: the first Lancashire canal', *Trans.Hist.Soc.Lancs. and Cheshire*, 100, 1948, 121-155.

Beare, T., *Hudson, The Education of an Engineer*, Edinburgh, 1901.

Becker, B. H., *Scientific London*, London, 1874; Frank Cass reprint, London, 1968.

Belliot, H. H., *University College London 1826-1926*, London, 1929.

Berridge, P. S. A., *Couplings to the Khyber*, David and Charles, Newton Abbot, 1969.

Berridge, P. S. A., *The Girder Bridge - After Brunel and Others*, Maxwell, London, 1969.

Bettenson, E. M., *The University of Newcastle upon Tyne*, University of Newcastle-on-Tyne, Newcastle, 1971.

Binnie, G. M., *Early Victorian Water Engineers*, Thomas Telford Ltd., London, 1981.

Blainey, G., *The Rush that Never Ended: A History of Australian Mining*, Melbourne, 1963.

Blainey, G., *The Tyranny of Distance: How distance shaped Australia's History*, Sun Books, 1966; Macmillan Australia, 1968.

Borthwick, A., *Yarrow and Company Limited: The First Hundred Years 1865-1965*, Yarrow and Co. Ltd., 1965.

Briggs, A., *Victorian People*, Odhams Press, London, 1954.

Brittain, Vera, *Women's Work in Modern England*, Noel Douglas, London, 1928.

Brock, W. H., 'The Japanese Connexion: engineering in Tokyo, London and Glasgow at the end of the nineteenth century', *British Journal of the History of Science*, 14, 3, 48, November 1981, 227-243.

Brooke, D., *The Railway Navvy: 'That Despicable Race of Men'*, David and Charles, Newton Abbot, 1983.

Buchanan, Brenda J., 'The Evolution of the English Turnpike Trusts: Lessons from a Case Study', *Econ.Hist. Review* 2nd series, 39, 1986, 223-243.

Buchanan, R. A., *Technology and Social Progress*, Pergamon, Oxford, 1965.

Buchanan, R. A., *Industrial Archaeology in Britain*, Pelican, Harmondsworth, 1972.

Buchanan, R. A., *History and Industrial Civilisation*, Macmillan, London, 1979

Buchanan, R. A., 'Gentlemen Engineers: The Making of a Profession', *Victorian Studies* 26, 4, Summer 1983, 407-429.

Buchanan, R. A., 'Institutional Proliferation in the British Engineering Profession, 1847-1914', *Econ.Hist.Review* 2nd series 38, 1, February 1985, 42-60.

Buchanan, R. A., 'The Rise of Scientific Engineering in Britain', *British Journal Hist.Sci.* 18, 2, 59, July 1985, 218-233.

Buchanan, R. A., 'The British Contribution to Australian Engineering', *Historical Studies*, Melbourne, 1983, 401-419.

Buchanan, R. A., 'The *Great Eastern* Controversy: A Comment', *Technology and Culture*, 24 January 1983, 98-106.

Buchanan, R. A., 'The Diaspora of British Engineering', *Technology and Culture*, 27, 3, July 1986, 501-524.

Buchanan, R. A., 'The British Canal Engineers: the men and their resources', Per Sörbom (ed.), *Transport Technology and Social Change*, Tekniska Museet Symposia, Stockholm, 1980, 67-89.

Buchanan, R. A., 'Steam and the Engineering Community in the Eighteenth Century', *Trans.Newcomen Soc.* 50, 1978-9, 193-202.

Buchanan. R. A., 'The Construction of the Floating Harbour in Bristol, 1804-1809', *Trans.Bristol and Glos.Arch.Soc.* 88, 1969, 184-204.

Buchanan, R. A., 'The Overseas Projects of I. K. Brunel', *Trans.Newcomen Society*, Presidential Address, 54, 1982-83, 145-166.

Buchanan, R. A. (with Doughty, M. W.), 'The choice of steam engine manufacturers by the British Admiralty, 1822-1852', *Mariner's Mirror*, 64, 4,1978, 327-347.

Buchanan, R. A. (with Jones, S.), 'The Balmoral Bridge of I. K. Brunel', *Industrial Archaeology Review*, 4, 3, Autumn 1980, 214-226.

Buchanan, R. A., 'From Trade School to University', in Walters, G. (ed.), *A Technological University - an experiment in Bath*, Bath, 1966, 12-26.

Buchanan, R. A., 'Engineering in the International Community', *Proc.16th Internat.Congress of Hist.Science*, Pt.B, 30-35, Bucharest, 1981.

Buchanan, R. A., 'Education or Training? The Dilemma of British Engineering in the Nineteenth Century', in Kranzberg, M. (ed), *Technological Education - Technological Style*, San Francisco, 1986, 69-73.

Buchanan, R. A., 'Engineers and government in nineteenth century Britain', in R. MacLeod (ed.), *Government and Expertise*, Cambridge University Press, 1988, 41-58.

Buckland, W., *The Life and Correspondence of William Buckland, DD, FRS* by his daughter, Mrs Gordon, London, 1894.

Burnham, J., *The Managerial Revolution*, Penguin, Harmondsworth, 1941.

Burton, A., *The Canal Builders*, Eyre Methuen, London, 1972.

Caff, W. R. M., *The History of the Development of the Steam Engine to the year 1850, with special reference to the work of West Country Engineers*, London University M.Sc. Dissertation, 1937.

Cameron, A. D., *The Caledonian Canal*, Terence Dalton Ltd., Lavenham, Suffolk, 1972.

Cameron, J. G. P., *A Short History of the Royal Indian Engineering College Coopers Hill*, Coopers Hill Society, 1960.

Campbell-Allen, D., and Davis, E. H. (eds), *The Profession of a Civil Engineer*, Sydney University Press, 1979.

Cantlie, K., *The Railways of China*, China Society, London, 1981.

Cantrell, J. A., *James Nasmyth and the Bridgewater Foundry*, Chetham Society, Manchester University Press, 1984.

Cardwell, D. S. L., *The Organisation of Science in England*, Heinemann, London, 1957.

Cardwell, D. S. L., *From Watt to Clausius: The Rise of Thermodynamics in the Early Industrial Age*, Heinemann, London, 1971.

Cardwell, D. S. L., *Technology, Science and History*, Heinemann, London, 1972.

Carr, L. H. A., *The History of the North-Western Centre of the Institution of Electrical Engineers*, Institution of Electrical Engineers, London, 1950.

Carr-Saunders, A. M., and Wilson, P. A., *The Professions*, Oxford, 1933.

Carter, E. C., *The Virginian Journal of Benjamin Henry Latrobe 1795-1798*, Maryland Historical Society, 1977.

Carus-Wilson, E. M., 'An Industrial Revolution of the Thirteenth Century', *Econ.Hist. Review*, 1st series, 11, 1941: reprinted in Carus-Wilson (ed.), *Essays in Economic History* Edward Arnold, London, 1, 1954, 41-60.

Catling, H., *The Spinning Mule*, David and Charles Library of Textile History, Newton Abbot, 1970.

Chapman, A. W., *The Story of a Modern University* (Firth College, Sheffield), Oxford, 1955.

Checkland, S. G., *The Rise of Industrial Society in England 1815-1885*, Longman, London, 1964.

Clark, E. K., *Kitsons of Leeds 1837-1937*, Loco.Pub.Co., London, n.d. but presumably 1937.

Clew, K. R., *The Kennet and Avon Canal*, David and Charles, Newton Abbot, 1968.

Close, Col. Sir Charles, *The Early Years of the Ordnance Survey*, Institution of Royal Engineers, London, 1926, and David and Charles, Newton Abbot, 1969.

Coe, W. E., *The Engineering Industry of the North of Ireland*, David and Charles, Newton Abbot, 1969.

Collier, D. A., *A comparative history of the development of the leading stationary steam engine manufacturers of Lancashire, c.1800-1939*, Unpublished Ph.D. Thesis, Manchester University, 1985.

Coleman, D. C., 'Gentlemen and Players', *Econ.Hist.Review* 2nd series, 26, 1, 1973, 92-116.

Corbett, A. H., *The Institution of Engineers Australia... 1919-1969*, Institution of Engineers, Australia, Sydney, 1973.

Corlett, E., *The Iron Ship: the History and Significance of Brunel's 'Great Britain'*, Moonraker Press, Bradford-on-Avon, 1974.

Cottle, B. and Sherborne, J. W., *The Life of a University*, Bristol, 1967.

Crump, W. B. (ed.), *The Leeds Woollen Industry 1780-1820*, The Thoresby Society, Leeds, 1931.

Cumming, D. A. and Moxham, G., *They built South Australia - Engineers, Technicians, Manufacturers, Contractors and their work*, Published by the authors, Adelaide, 1986.

Daniels, G. W., *The Early English Cotton Industry*, Manchester University Press, 1920.

Darby, H. C., *The Draining of the Fens*, Cambridge University Press, 1st edition 1940, second edition 1956, reprinted 1968.

Darby, H. C. (ed.), *An Historical Geography of England before AD 1800*, Cambridge University Press, 1936, reprinted 1948.

David, P. A., 'The landscape and the machine: technical interrelatedness, land tenure and the mechanization of the corn harvest in Victorian Britain', in McCloskey, D. N. (ed.) *Essays on a Mature Economy: Britain after 1840*, Methuen, London, 1971, 145-214.

Davies, A. S., 'The Coalbrookdale Company and the Newcomen Engine, 1717-1769', *Trans. Newcomen Soc.* 20, 1939-40, 45-48.

Dewhurst, P. C., 'The Fairlie Locomotive', *Trans.Newcomen Society*, Pt.I - 34,1961-2, 105-132; Pt.II - 39,1966-7,1-34.

Dickinson, H. W., *A Short History of the Steam Engine*, Cambridge, 1938, and Frank Cass, London, 1963.

Dickinson, H. W. and Gomme, A. A., 'Some British contributions to continental technology (1600-1850)', *Actes du VIe Congrès International d'Histoire des Sciences* Amsterdam 1950, 1, 307-323.

Donald, M. B., *Elizabethan Monopolies - A History of the Company of Mineral and Battery Works from 1565 to 1604*, Oliver and Boyd, Edinburgh and London, 1961.

Dugan, J., *The Great Iron Ship*, Hamish Hamilton, London, 1953.

Dumbell, S., *The University of Liverpool 1903-53*, University of Liverpool, Liverpool, 1953.

Emmerson, G. S., *Engineering Education: A Social History*, David and Charles, Newton Abbot, 1973.

Engineers and Officials: an historical account of 'Health of Towns Works' (between 1838 and 1856) in London and the Provinces, London, 1856.

The Engineers and Machinist's Assistant..., new and improved ed., Blackie, Edinburgh, 1850, 2 vols.

Ewing, J. A., *The University Training of Engineers*, Cambridge, 1891.

Eyles, Joan M., 'William Smith: Some Aspects of His Life and Work', in Schneer, C. J. (ed.) *Toward a History of Geology* MIT Press 1969, 142-158: see also article in *Dictionary of Scientific Biography* New York, 1975, 12.

Fairbairn, Sir William, *Treatise on Mills and Millwork*, Longman, London, 1861.

Fieldhouse, D. K., *Economics and Empire 1830-1914*, Weidenfeld, London, 1973.

Finniston, Sir Montague, *Engineering our Future: Report of the Committee of Inquiry into the Engineering Profession*, HMSO, London, Cmnd.7794, 1980.

Floud, R., *The British Machine Tool Industry*, 1850-1914, Cambridge, 1976.

Fox, R. and Weisz, G., *The Organization of Science and Technology in France 1808-1914*, Cambridge University Press, 1980.

Fraser, D., 'The Politics of Leeds Water', *Pubs.Thoresby Soc.*, 53; The Thoresby Miscellany, 15, Leeds, 1973.

Gallman, R. E. (ed.), *Recent Developments in the study of Business and Economic History*, JAI Press, Greenwich, Connecticut, 1977.

Gerstl, J. E. and Hutton, S. P., *Engineers: the anatomy of a profession - a study of mechanical engineers in Britain*, Tavistock Press, London, 1966.

Glover, Ian A. and Kelly, Michael P., *Engineers in Britain: A Sociological Study of the Engineering Dimension*, Allen and Unwin, London, 1987.

Gosden, P. H. J. H. and Taylor, A. J., *Studies in the History of a University 1874-1974*, E. J. Arnold, Leeds, 1975.

Grant, Sir Allan, *Steel and Ships: The History of John Brown's*, Michael Joseph, London, 1950.

Green, E., *Debtors to their Profession - A History of the Institute of Bankers*, Methuen, London, 1979.

Habakkuk, H. J., *American and British Technology in the Nineteenth Century*, Cambridge, 1962.

Hadfield, C., *British Canals: An Illustrated History*, David and Charles, Newton Abbot, 1st edition 1950, 5th edition 1974.

Hadfield, C., *The Canal Age*, David and Charles, Newton Abbot, 1968.

Hague, D. B. and Christie, R., *Lighthouses: their architecture, history, and archaeology*, Gomer Press, Llandysul, Dyfed, 1975.

Hamilton, S. B., 'Continental influences on British civil engineering to 1800', *Archives Internationales d'Histoire des Sciences*, 2, 1958, 347-55.

Harley, C. K., 'The shift from sailing ships to steamships, 1850-1890: a study in technological change and its diffusion', in McCloskey, D. N. (ed.) *Essays in a Mature Economy: Britain after 1840*, Methuen, London, 1971, 215-237.

Harris, H., *The Industrial Archaeology of the Peak District*, David and Charles, Newton Abbot, 1971.

Harris, J. R., 'The early steam engine on Merseyside', *Trans.Hist.Soc.Lancs.and Cheshire*, 106, 1954, 109-116.

Harris, T. R., 'Engineering in Cornwall before 1775', *Trans.Newcomen Soc.* 25, 1945-47, 111-122.

Hartwell, R. M. (ed.), *The Causes of the Industrial Revolution in England*, Methuen, London, 1967.

Hatchett, C., *The Hatchett Diary: A tour through the counties of England and Scotland in 1796 visiting their mines and manufactories*, Edited with introduction by Arthur Raistrick, D. Bradford Barton, Truro, 1967.

Haut, F. J. G., 'The Centenary of the Semmering Railway and its Locomotives', *Trans.Newcomen Society*, 27, 1949-51, 19-29.

Hawke, G. R., *Railways and Economic Growth*, Oxford, 1970

Head, Sir Francis Bond, *Stokers and Pokers, or the London and North-Western Railway...*, London, 1849, reprinted Frank Cass, London, 1968.

Hearnshaw, F. J. C., *The Centenary History of King's College London, 1828-1928*, G. G. Harrop, London, 1929.

Henderson, W. O., *The Industrialization of Europe 1780-1914*, Thames and Hudson, London, 1969.

Hilken, T. J. N., *Engineering at Cambridge University 1783-1965*, Cambridge University Press, 1967.

Hills, R. L., *Machines Mills and Uncountable Costly Necessities: A short history of the drainage of the Fens*, Goose and Son, Norwich, 1967.

Hills, R. L., *Power in the Industrial Revolution*, Manchester University Press, 1970.

Hindle, B., *Emulation and Invention*, New York University Press, New York, 1981.

Hobsbawm, E. J., *Industry and Empire*, Pelican, Harmondsworth, 1968.

Hogg, J. (ed.), *Fortunes made in Business*, London, 1884.

Hooper, W. T., 'Perran Foundry and its Story', *Royal Cornwall Polytechnic Soc.* n.s. 9, 3, 1939, 62-89.

Huelin, G., *King's College London 1828-1978*, University of London, King's College, London, 1978.

Hughes, E., 'The first steam engines in the Durham Coalfield', *Archaelogia Aeliana* 27, 1949, 29-45.

Hughes, E., 'The professions in the eighteenth century', *Durham University Journal* new series, 13, 1952, 46-55.

Hughes, Thomas P., *Networks of Power - Electrification in Western Society 1880-1930*, Johns Hopkins University Press, Baltimore and London, 1983.

Hurd, J., 'Railways' in *The Cambridge Economic History of India*, 2, c.1757-c.1970, Cambridge, 1980.

Imperial Gazetteer of India - The Indian Empire, III Economic, Oxford, 1908.

Inkster, I., 'The development of a scientific community in Sheffield, 1790-1850', *Trans.Hunter Archaeological Soc.* 10, 2, 1973, 99-131.

Jefferys, J. B., *The Story of the Engineers 1800-1945*, Lawrence and Wishart, London, 1945.

Jenkin, F., *A lecture on the education of civil and mechanical engineers in Great Britain and abroad*, Edmonston and Douglas, Edinburgh, 1868.

Jenkins, D. T., *The West Riding Wool Textile Industry 1770-1835: A Study of Fixed Capital Formation*, Pasold, Edington, Wilts, 1975.

Jenkins, R., *The Collected Papers of Rhys Jenkins, M.I.Mech.E.*, Cambridge, 1936.

Jensen, M., *Civil Engineering around 1700*, Danish Technical Press, Copenhagen, 1969.

Jeremy, D. J., *Transatlantic Industrial Revolution: The Diffusion of Textile Technologies between Britain and America, 1790-1830s*, MIT Press and Blackwells, Oxford, 1981.

Jespersen, A., *A Preliminary Account of the Development of the Gearing in Watermills in Western Europe*, Denmark, 1953.

Jewkes, J., Sawers, D., and Stillerman, R., *The Sources of Invention*, Macmillan, 2nd edition 1969.

Jones, E. L., *Agriculture and the Industrial Revolution*, Basil Blackwell, Oxford, 1974.

Jones, L. J., 'The early history of mechanical harvesting', A. R. Hall and N. Smith (eds) *History of Technology*, 4, 1979, 101-148.

Jones, L. J., 'Wind, water and muscle-powered flour-mills in early South Australia', *Trans.Newcomen Society*, 53, 1981-82, 97-118.

Kargon, R. H., *Science in Victorian Manchester: Enterprise and Expertise*, Manchester University Press, 1977.

Kaye, B., *The Development of the Architectural Profession in Britain*, London, 1960.

Kirby, R. S. and Laurson, P. G., *The Early Years of Modern Civil Engineering*, New Haven, Yale University Press, USA; also London and Oxford, 1932.

Landes, D. S., *The Unbound Prometheus- Technological Change and Industrial Development in Western Europe from 1750 to the Present*, Cambridge University Press, 1969.

Lazonick, W., 'Industrial relations and technical change: the case of the self-acting mule', *Cambridge Jnl. of Econ.* 3, 1979, 231-262.

Lean, T., *Historical Statement of the Improvements made in the duty performed by the Steam Engines in Cornwall, 1839*, reprinted by Bradford Barton, Truro, 1969.

Lee, C. H., *A cotton enterprise 1795-1840: A history of M'Connel and Kennedy*, Manchester University Press, 1972.

Leggett, R. F., 'The Jones Falls Dam on the Rideau Canal', *Trans.Newcomen Society*, 31, 1957-59, 205-218.

Leleux, S. A., *Brotherhoods, Engineers*, David and Charles, Dawlish, 1965.

Lewis, R. and Maude, A., *Professional People*, Phoenix House, London, 1952.

Lindqvist, S., *Technology on Trial: The Introduction of Steam Power Technology into Sweden, 1715-1736*, Uppsala and Stockholm, 1984.

Lindqvist, S., 'The work of Martin Triewald in England', *Trans.Newcomen Society*, 50, 1978-79, 165-172.

Linge, G. J. R., *Industrial Awakening: A Geography of Australian Manufacturing 1788-1890*, Australian National University, Canberra, 1979.

Lloyd, B. E., *The Education of Professional Engineers in Australia*, Assoc.of Professional Engineers Australia, Melbourne, 1968.

Lloyd, B. E., *In Search of Identity: Engineering in Australia 1788-1988*, University Melbourne, unpublished Ph.D. thesis, September 1988.

MacDonagh, O., *A Pattern of Government Growth 1800-60*, MacGibbon and Kee, London, 1961.

McCloskey, D. N. (ed.), *Essays on a Mature Economy: Britain after 1840*, Methuen, London, 1971.

Mackie, J. D., *The University of Glasgow 1451-1951*, Jackson, Son and Co., Glasgow, 1954.

MacLeod, R. M., 'The Alkali Acts Administration 1863-84', *Victorian Studies*, 9, 2, 1965, 85-112.

MacLeod, R. M. (ed.), *Government and Expertise: specialists, administrators and professionals, 1860-1919*, Cambridge University Press, 1988.

Malet, H., *The Canal Duke*, Phoenix House, London, 1961.

Mantoux, P., *The Industrial Revolution in the Eighteenth Century*, English translation, London, 1928; Jonathan Cape revised edition, London, 1961.

Maré, E. de, *Swedish Cross Cut: A Book on the Götha Canal*, Malmo, Sweden, 1964.

Marr, D. B., *Recollections and Memoirs of the Belfast Association of Engineers*, Belfast, 1967.

Marsden, B.M., *Pioneers of Prehistory: Leaders and Landmarks in English Archaeology*, G. W. and A. Hesketh, Ormskirk and Northridge, 1984.

Marsh, J. O., 'The engineering institutions and the public recognition of British engineers', *Internat.Jnl.Mech.Eng.Education*, 16, 2, April 1988, 119-127.

Marshall, D. W., *The British Military Engineers 1741-1783: A study of organization, social origin, and cartography*, Unpublished Ph.D dissertation, University of Michigan, 1976.

Marx, K., *Capital*, 1867; Penguin ed. 1, Harmondsworth, 1976.

Mathias, P., *The First Industrial Nation*, Methuen, London, 1969.

Merritt, R. H., *Engineering in American Society 1850-1875*, University of Kentucky, 1969.

Metropolitan Water Board, *London's Water Supply 1903-1953*, MWB, London, 1953.

Middlemas, R. K., *The Master Builders*, Hutchinsons, London, 1963.

More, C., *Skill and the English Working Class 1870-1914*, Croom Helm , London, 1980.

Morrell, J. and Thackray, A., *Gentlemen of Science: Early Years of the British Association for the Advancement of Science*, Clarendon Press, Oxford, 1981.

Morris, R. J., 'The Rise of James Kitson: Trades Union and Mechanics Institution, Leeds, 1826-1851', *The Thoresby Miscellany* 15, Leeds, 1973, 179-200.

Mullineux, F., *The Duke of Bridgewater's Canal*, Eccles and District History Society, 1959.

Musson, A. E. and Robinson, E., *Science and Technology in the Industrial Revolution*, Manchester University Press, 1969.

Neale, R. S., *Class and Ideology in the Nineteenth Century*, Routledge, London, 1972.

Namier, L. B., *The structure of politics at the accession of George III*, London, 1929.

Nef, J. V., 'The Progress of Technology and the Growth of Large-Scale Industry in Great Britain, 1540-1640': *Econ.Hist. Review*, 1st series, 5, 1934; reprinted in Carus-Wilson (ed.), *Essays in Economic History*, Edward Arnold, London, 1,1954, 88-107.

Oakley, C. A., 'The mechanical engineering industry of Clydeside - its origins and development', *Trans.Inst.Engs.and Shipbuilders Scotland*, 89, 1945-46, 9-39.

Oriel, J. A., 'The technological awakening', *Trans.Inst.Chem.Engs*. 34, 1956, 113-116.

Ormond, R., *Sir Edwin Landseer*, Exhibition Catalogue, London, 1981.

Pacey, A., *The Maze of Ingenuity*, Allen Lane, London, 1974.

Pacey, A., *The Culture of Technology*, Basil Blackwell, Oxford, 1983.

Parris, H., *Government and the Railways in Nineteenth-Century Britain*, Routledge, London, 1965.

Parris, H., *Constitutional Bureaucracy: The development of British central administration since the eighteenth century*, Allen and Unwin, London, 1969.

Parris, H., 'The Nineteenth-century Revolution in Government: a Reappraisal reappraised', *Historical Journal* 3,1,1960, 17-37.

Parris, H., 'A Civil Servant's Diary, 1841-46', *Public Administration*, 38, 1960, 369-380.

Parris, H., 'Pasley's Diary: a neglected source of railway history', *Jnl.Transport Hist.*, 6, 1963-4, 14-23.

Parsons, R. H., *The Early Days of the Power Station Industry*, Cambridge University Press, 1940.

Patterson, A. T., *The University of Southampton*, University of Southampton, Southampton, 1962.

Pendred, L. St L., *British Engineering Societies*, Longman, for British Council, London, 1947.

Perkin, H., *The Origins of Modern English Society 1780-1880*, Routledge, London, 1969.

Perrucci, R. and Gerstl, J. E. (eds), *The Engineers and the Social System*, Wiley, New York, 1969.

Pollard, S., *The Genesis of Modern Management*, London 1965 and Penguin, Harmondsworth, 1968.

Porter, C. T., *Engineering Reminiscences*, contributions to *Power* and *American Machinist* republished New York, 1908.

Preece, C., 'The Durham Engineer Students of 1838', *Trans.Arch.and Arch.Soc.of Durham and Northumberland*, n.s.6,1982, 71-74.

Pursell, Carroll W., *Early Stationary Steam Engines in America: A Study in the Migration of a Technology*, Smithsonian Institution Press, Washington D.C., 1969.

Purser, J., 'Note on the Engineering School since its Foundation' (TCD), *Hermathena*, 58, November 1941, 53-56.

Raistrick, A., *Quakers in Science and Industry*, Bannisdale Press, 1950; David and Charles, Newton Abbot, 1968.

Raistrick, A., 'The Steam Engine on Tyneside, 1715-1778', *Trans.Newcomen Society*, 17, 1936-7, 131-164.

Reader, W. J., *Professional Men: the rise of the professional classes in nineteenth-century England*, Weidenfeld, London, 1966.

Reader, W. J., '"At the head of all the new professions": the engineer in Victorian society' in McKendrick, N. and Outhwaite, R. B. (eds.), *Business Life and Public Policy Essays in honour of D. C. Coleman*, Cambridge University Press, 1986.

Reader, W. J., *A History of the Institution of Electrical Engineers 1871-1971*, Peter Perigrinus, London, 1987.Redding, C., Yesterday and Today, 3 vols. London, 1863.

Reed, B., *Crewe Locomotive Works and its Men*, David and Charles, Newton Abbot, 1982.

Reed, M. C. (ed.), *Railways in the Victorian Economy*, David and Charles, Newton Abbot, 1969.

Reed, M. C., *Investment in Railways in Britain, 1820-1844: A study in the development of the capital market*, Oxford, 1975.

Rennison, R. W., *Water to Tyneside: A History of the Newcastle and Gateshead Water Company*, Newcastle and Gateshead Water Co., Gateshead, 1979.

Richards, J. M., *The Functional Tradition in Early Industrial Buildings*, Architectural Press, London, 1958.

Rimmer, W. G., *Marshalls of Leeds: Flax-Spinners 1788-1886*, Cambridge, 1960.

Ritchie-Noakes, Nancy, *Liverpool's Historic Waterfront - The World's First Mercantile Dock System*, HMSO for RCHME, London, 1984.

Robson, R. (ed.), *Ideas and Institutions of Victorian England: Essays in honour of George Kitson Clark*, Bell, London, 1967.

Roll, E., *An Early Experiment in Industrial Organisation, being a history of the firm of Boulton and Watt, 1775-1805*, Longman, London, 1930.

Rolt, L. T. C., *Tools for the Job - A Short History of Machine Tools*, Batsford, London, 1965.

Rolt, L. T. C., *Victorian Engineering*, Allen Lane, Harmondsworth, 1970.

Rosenberg, N., and Vincenti, W. G., *The Britannia Bridge: The Generation and Diffusion of Technological Knowledge*, MIT Press, Cambridge Mass., 1978.

Rostow, W. W., *The Process of Economic Growth*, Oxford, 1960

Rowe, J. H., 'The early history of Hayle Foundry 1770-1833', *Royal Cornwall Polytechnic Soc.* n.s.8, 3, 1936, 40-49.

Rudwick, Martin J. S., *The Great Devonian Controversy: the shaping of scientific knowledge among gentlemanly specialists*, Chicago University Press, 1985.

Sampson, A., *Anatomy of Britain*, Hodder and Stoughton, London, 1962.

Sanderson, M., *The Universities and British Industry 1850-1970*, Routledge, London, 1972.

Salis, H.R. de, *A Chronology of Inland Navigation in Great Britain*, London, 1897.

Saul, S. B., *The Myth of the Great Depression 1873-1896*, Macmillan, London, 1969.

Saul, S. B. (ed.), *Technological Change: the United States and Britain in the Nineteenth Century*, Methuen, London, 1970.

Saul, S. B., 'The Machine Tool Industry in Britain to 1914', *Business History* 10, 1968, 22-43.

Saul, S. B., 'The Market and the Development of the Mechanical Engineering Industries in Britain, 1860-1914', in Saul, S. B. (ed.), *Technological Change: the United States and Britain in the Nineteenth Century*, Methuen, London, 1970, 141-170; first pub. *Econ.Hist.Review*, 2nd series, 20, 1967.

Saul, S. B., 'The Nature and Diffusion of Technology', in Youngson, A. J. (ed.), *Economic Development in the Long Run*, Unwin, London, 1972, 36-61.

Schofield, R. E., *The Lunar Society of Birmingham: A Social History of Provincial Science and Industry in Eighteenth Century England*, Oxford, 1963.

Sebestik, J., 'The Rise of the Technological Science', *History and Technology*, 1, 1, 1983, 25-43.

Shapin, S. and Thackray, A., 'Prosopography as a research tool in history of science: the British scientific community 1700-1900', *History of Science* 12, 1974, 1-28.

Shimmin, A. N., *The University of Leeds*, Cambridge, 1954.

Simmons, J., *Parish and Empire: Studies and Sketches*, Collins, London, 1952..

Simmons, J., *The Railways of Britain*, Routledge, London, 1961; second ed. Macmillan, London, 1968.

Simmons, J. (ed.), *The Birth of the Great Western Railway, Extracts from the Diary and Correspondence of George Henry Gibbs*, Adams and Dart, Bath, 1971.

Simms, F. W. (ed.), *Public Works of Great Britain*, John Weale, Architectural Library, London, 1838.

Sinclair, B., 'Canadian Technology: British Traditions and American Influences', *Technology and Culture*, 20, 1, 1969, 108-123.

Sinclair, B., *A Centennial History of the American Society of Mechanical Engineers 1880-1980*, ASME, University of Toronto Press, Toronto, 1980.

Skelton, R. A., 'The Origins of the Ordnance Survey of Great Britain', *Geographical Journal*, 128, 1962, 415-430.

Skempton, A. W., 'Engineering on the Thames Navigation', *Trans.Newcomen Society*, 55, 1983-84, 153-176.

Small, J., 'Glasgow University's contribution to engineering progress', *Glasgow University Engineering Soc.Year Book*, 1954.

Smith, N. (ed.), *History of Technology*, Eleven vols., Mansell, London, 1976-1986.

Smith, N., *Man and Water: A History of Hydro-Technology*, Peter Davies, London, 1976.

Sörbom, P. (ed.), *Transport Technology and Social Change*, Tekniska Museet Symposia No.2, Stockholm, 1980.

Southey, R., *Espriella's Letters from England*, London, 1807.

Spring, D., *The English Landed Estate in the Nineteenth Century*, Johns Hopkins, Baltimore, 1963.

Stone, I., *Canal Irrigation in British India: Perspectives on technological change in a peasant economy*, Cambridge, 1984.

Stone, L., 'Prosopography', *Daedalus*, Winter 1971, 46-79.

Sussman, H. L., *Victorians and the Machine: The Literary Response to Technology*, Harvard University Press, Cambridge Mass., 1968.

Sutherland, G. (ed.), *Studies in the growth of nineteenth-century government*, Routledge, London, 1972.

Tann, J., *Boulton and Watt's organisation of steam engine production before the opening of the Soho Foundry*, University of Aston, Birmingham, 1978.

Tann, J., *The Development of the Factory*, Cornmarket, London, 1970.

Tann, J. and Breckin, M. J., *Fixed capital formation in steam power: a case study of the Boulton and Watt engine*, University of Aston, Birmingham, 1978.

Tann, J., 'Fuel saving in the process industries during the Industrial Revolution: a study in technological diffusion', *Business History* 15, 2, 1973, 149-159.

Thackray, A., 'Science and Technology in the Industrial Revolution', *History of Science*, 9, 1970, 76-89.

Taylor, W., *The Military Roads of Scotland*, David and Charles, Newton Abbot, 1976.

Tew, D. H., 'Canal Lifts and Inclines', *Trans.Newcomen Society*, 28, 1951-53, 35-58.

Thompson, F. M. L., *Chartered Surveyors: the growth of a profession*, Routledge, London, 1968.

Thompson, F. M. L., *English Landed Society in the Nineteenth Century*, Routledge, London, 1963.

Thompson, Joseph, *The Owens College, its foundation and growth*, J. E. Cornish, Manchester, 1886.

Trebilock, C., *The Vickers Brothers: Armaments and Enterprise 1854-1914*, Europa, London, 1977.

Trevithick, R. F., 'Locomotive Building in Japan', *Proc.Inst.Mech.Engs.* 1895, 298-307.

Trollope, A., *The Claverings*, London, 1867.

Torstendahl, R., *Dispersion of Engineers in a Transitional Society*, Studia Historica Upsaliensia, 73, Uppsala, Sweden, 1975.

Toynbee, A., *Lectures on the Industrial Revolution in England*, London, 1884.

Tunzelmann, G. N. von, 'Technological Diffusion during the Industrial Revolution', in R. M. Hartwell (ed.), *The Industrial Revolution*, Oxford, 1970.

Tunzelmann, G. N. von, *Some Economic Aspects of the Diffusion of Steam Power in the British Isles to 1856, with special reference to Textile Industries*, Oxford, unpublished D.Phil. thesis, 1974.

Tunzelmann, G. N. von, *Steam Power and British Industrialization to 1860*, Clarendon Press, Oxford, 1978.

Tupling, G. H., 'The early metal trades and the beginnings of engineering in Lancashire', *Trans.Lancs.and Ches.Antiquarian Soc.*,61, 1949, 1-34.

Turner, A. L. (ed.), *History of the University of Edinburgh 1883-1933*, University of Edinburgh. Edinburgh, 1933.

Vincent, E. W. and Hinton, P., *The University of Birmingham: its history and significance*, Cornish Bros., Birmingham, 1947.

Walker, C., *Joseph Locke*, Shire Publications, Aylesbury, 1975.

Walker, R. J. B., *Old Westminster Bridge - The Bridge of Fools*, David and Charles, Newton Abbot, 1979.

Ward, J. R., *The Finance of Canal Building in eighteenth-century England*, Oxford, 1974.

Warren, J. G. H., *A Century of Locomotive Building by Robert Stephenson and Co. 1823-1923*, Newcastle-upon-Tyne, 1923.

Waterhouse, R. E., *A Hundred Years of Engineering Craftsmanship*, Tangyes, Cornwall Works, Smethwick, Birmingham, 1957.

Watson, Garth, *The Civils: The Story of the Institution of Civil Engineers*, Thomas Telford, London, 1988.

Westwood, J. N., *Railways of India*, David and Charles, Newton Abbot, 1974.

White, H., 'Fossets': *A record of two centuries of engineering*, Fawcett Preston and Co. Ltd., 1958.

Whiting, C. E., *The University of Durham 1832-1932*, Sheldon Press, London, 1932.

Wiener, M. J., *English Culture and the Decline of the Industrial Spirit 1850-1980*, Cambridge University Press, 1981.

Willans, T. S., *River Navigation in England, 1600-1750*, second edition, London, 1964.

Williams, A. F., 'Bristol Port Plans and Improvement Schemes of the Eighteenth Century', *Trans.Bristol and Glos.Arch.Soc.* 81, 1962

Wilson, C. and Reader, W., *Men and Machines: A History of D. Napier and Son, Engineers, Ltd., 1808-1958*, Weidenfeld, London, 1958.

Wood, A. C., *The History of University College Nottingham*, Oxford, 1953.

Wood, Sir Henry Trueman, *A History of the Royal Society of Arts*, John Murray, London, 1913.

Wootton, Barbara, *The Social Foundations of Wage Policy*, Allen and Unwin, London, 1955.

Appendix

Table I

Membership of the British Engineering Institutions, 1850-1914

Institution	Founded	1850	1860	1870	1880	1890	1900	1910	1914	Notes
Institution of Civil Engineers	1818	664	894	1589	2960	4739	6303	8843	9194	Royal Charter 1828
Institution of Mechanical Engineers	1847	201	c.400	957	1507	2805	3165	5583	6400	Incorp. 1878; Royal Charter 1929
Institution of Naval Architects	1860	-	365 (1861)	480	510	950	1500	1990	2100	Charter 1910; 'Royal' 1959
Institution of Gas Engineers	1863	-	-	c.350	c.700	c.700	c.700	829	875	Incorp. 1890, 1902; Charter 1929
Royal Aeronautical Society	1866	-	-	c.100	c.90	c.65	c.40	291	348 (1912)	'Royal' 1918; Women - 1898
Iron and Steel Institute	1869	-	-	300	1100	1590	1600	2200	2100	Royal Charter 1899; Metals Society 1974
Institution of Electrical Engineers	1871	-	-	352 (1871)	c.1000	2100	c.4000	6218	7045	Incorp. 1883; 'Soc. Teleg. Engs.' to 1888; Charter 1921
Institution of Municipal Engineers	1873	-	-	-	180	403	966	1257	1583	Charter 1890
Institution of Marine Engineers	1889	-	-	-	-	452	938	1228	1467	Charter 1933
Institution of Mining Engineers	1889	-	-	-	-	1239	2482	3254	3277	Charter 1918
Institution of Mining and Metallurgy (I.M.M.)	1892	-	-	-	-	-	615	1902	2372	Charter 1915
Institution of Public Health Engineers	1895	-	-	-	-	-	c.500	591	635	Incorp. 1916; 'Sanitary Engs.' to 1954
Institution of Water Engineers	1896	-	-	-	-	-	c.200	376	422	Incorp. 1911
Institution of Heating and Ventilating Engineers	1897	-	-	-	-	-	c.180	278	476	Incorp. 1950?
Institution of Automobile Engineers	1906	-	-	-	-	-	-	c.530	c.900	Royal Charter 1938
Institution of Structural Engineers	1908	-	-	-	-	-	-	850	1006 (1913)	Incorp. 1909; Royal Charter 1934
Institution of Locomotive Engineers	1911	-	-	-	-	-	-	52 (1912)	175 (1912)	Incorp. 1915
Totals	-	865	1,659	4,128	8,047	15,043	23,189	36,272	40,375	

Sources: *Proceedings* of the Institutions, extrapolated from graphs and other data where necessary, as marked by prefix 'c.'

Table II

Regional Professional Organisations of Engineers in Britain: Summary of the Main Institutions Since 1850

Title at Foundation	Developments 1850-1870	Developments 1870-1914	Developments since 1914
1. Institutions promoted by Mining Engineers			
North of England Institute of Mining Engineers	Fd. Newcastle-upon-Tyne 1852 'and mechanical' added 1866. 1853 = 80	Royal Charter 1876. Federated Institution of Mining Engineers 1889. 1876 = c.800	1981 = 540
South Wales Institute of Engineers	Fd. Merthyr Tydfil 1857.	Cardiff 1878. Royal Charter 1881	affl. Institution Mining Engineers, 1952. 1957 = 911
Chesterfield and Derbys. Institute of Mining Civil and Mechanical Engineers	Fd. Chesterfield 1871. 1871 = 118	Federated Institution of Mining Engineers 1889. 1901 = 375	Midland Counties Institution of Engineers since 1901. 1973 = 598
Miners' Association of Cornwall and Devonshire	Fd. Cornwall 1859. 1868 = 140	amalg. Mining Institute of Cornwall 1885: Mining Assoc. and Institute; 1887 = 258	No activities after 1895. Cornish Institute of Mining Mech. and Metal Engs. 1913-1930
2. Institutions promoted by Mechanical Engineers			
Leeds Association of Foremen Engineers and Draughtsmen	Fd. Leeds 1865	Leeds Association of Engineers 1890	Link with Keighley Association of Engineers (fd. 1900) 1921 = 382
Manchester Association of Employers, Foremen and Draughtsmen	Fd. Manchester 1856. 1857 = 12	Manchester Association of Engineers, 1885. 1878 = 178	1979 = 234
Birmingham and District Foremen and Draughtsmen's Mechanical Association		Fd. Birmingham 1889 Birmingham Association of Mechanical engineers 1891. 1905 = 370	No activities after 1942 1928 = 324

Glasgow and West of Scotland Association of Foremen Engineers and Draughtsmen	Fd. Glasgow 1898	West of Scotland Association of Foremen Engineers 1939. 1932 = 586. No activities since 1945.
Sheffield Society of Engineers and Metallurgists	Fd. Sheffield 1894. Amalg. Sheffield Society of Engineers (1888) and Sheff. Metallurgical Soc. (1890) 1914 = c.450	No activities after 1949

3. Institutions promoted by Civil Engineers and shipbuilders

Institution of Engineers in Scotland	Fd. Glasgow 1857: amalg. with Scottish Shipbuilders Association (fd. 1860) in 1866; 1858 = 127	Incorporated 1871: Inst. of Engineers and Shipbuilders in Scotland. 1871 = c.400	1914 = 1677; 1972 = 1063
Manchester Institution of Engineers	Fd. Manchester 1867; No activities after 1868.		
North East Coast Institution of Engineers and Shipbuilders	Fd. Newcastle-upon-Tyne 1884; 1885 = 452	Incorporated 1914 (Royal Patronage) 1981 = 1400	
Liverpool Engineering Society	Fd. Liverpool 1875 1885 = 110	Incorporated 1934 No activities after 1968. 1925 = 886; 1959 = 510	

Sources: Transactions and papers of regional organisations and official accounts as referred to in the accompanying text and footnotes.

Index